SCHUMACHER ON ENERGY

Speeches and Writings of
E F SCHUMACHER
edited by Geoffrey Kirk

WITHDRAWN

ABACUS

ABACUS edition published in 1983
by Sphere Books Ltd
30–32 Gray's Inn Road, London WC1X 8JL

First published in Great Britain by
Jonathan Cape 1982
Copyright © text 1982 by the Estate of E F Schumacher
Copyright © Introduction and linking material 1982 by Geoffrey Kirk

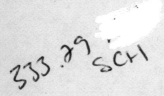
Reproduced, printed and bound in Great Britain by
Hazell Watson & Viney Ltd, Aylesbury, Bucks

E F Schumacher was Economic Adviser of the National Coal Board from
1950 to 1970. German born, he first came to England in the 1930s as a
Rhodes Scholar to study economics at New College, Oxford, and later
taught economics at Columbia University, New York. He served as
Economic Adviser with the British Control Commission in Germany
from 1946 to 1950. His advice on the problems of rural development was
sought by many overseas governments, and in 1974 he was awarded the
CBE. Dr Schumacher died in 1977. He was the author of SMALL IS
BEAUTIFUL* and A GUIDE FOR THE PERPLEXED*.

* Also available in Abacus

For Vreni
and the children

There is no substitute for energy; the whole edifice of modern life is built upon it. Although energy can be bought and sold like any other commodity, it is not 'just another commodity', but the precondition of all commodities, a basic factor equally with air, water and earth.

(From E F Schumacher, 'Energy supplies – the need for conservations', Energy International, September 1964)

Contents

Acknowledgments

George McRobie, author of *Small is Possible* and Fritz Schumacher's successor at the Intermediate Technology Development Group, first as Director, then as Chairman, suggested I should edit this book: I am also indebted to him and to David Brandrick for their advice and comments.

Mrs Verena Schumacher gave me permission to include extracts from her husband's papers and the National Coal Board allowed me to use material the copyright of which they own.

I am also indebted to the following for permission to reproduce copyright material: *Ambio* for chapter 1, extract 11 and chapter 2, extract 7; Blond & Briggs for chapter 4, extract 13, chapter 5, extract 10 and chapter 6, extract 7; Cambridge University Press (*Economic Journal*) for chapter 3, extract 1; Energy International for chapter 1, extract 1; the National Westminster Bank for chapter 5, extract 4; and *The Times* for chapter 3, extract 5.

I was helped by a host of present and former NCB colleagues, including Lord Robens, Dr Joe Gibson, Michael Parker, John Beamish-Crooke, Richard Ormerod, Brian Youngman, Brian Simpson, Joe McDonnell, Francis Gysin, John Scott, Ken Harding, Pam Evans, Eileen Sayer, and Betty Ramsbotham and Philip Toms of the Hobart House Library.

I want also to thank Peter Oppenheimer of Christ Church, Oxford, Satish Kumar of Resurgence, John Davis and Walter Marsden, for their kindness, which took many forms.

Articles published abroad were traced by Professor F.L. Wilke of the Technical University of Clausthal, W.F. Riester, and my German and French colleagues, Dr Alfred Plitzko and Jean-Pierre Rousselot.

The manuscript was typed for me by Beryl Browne.

G.K.

Editor's Introduction

This book brings together a representative selection of E.F. Schumacher's speeches and writings over a period of nearly twenty-five years on the subject of energy. The analysis he presented at the time, and the remedies he proposed, are still valid: the dangers he warned about have not been removed.

Also included are his views on management and public ownership, his thinking on which is particularly valuable since he worked for twenty years in the first major industry to be nationalised in Britain.

In each of his four careers — in social economics, energy, intermediate technology and writing — Schumacher achieved such distinction and recognition as would have contented most men who had followed only one of those activities. In each he made a contribution to human wellbeing that was not confined to one nation or, perhaps, to the present age. As some of the lectures show, he would undoubtedly have been an outstanding university teacher if he had followed an academic career — as he once intended.

It was typical of him, with his respect for the accumulated wisdom of mankind, that he should quote a Papal Encyclical in support of his case for the greatest possible devolution of managerial responsibility. The interest he developed in Buddhism when he was seconded to advise the Burmese Government led him to favour non-violent economic policies for the rest of his life; his final spiritual home, however, was in the Roman Catholic faith which he embraced wholeheartedly after, as he used to say, having enjoyed a long flirtation with it. His religious thinking is set out in *A Guide for the Perplexed*[1] — which he regarded as his most important book.

When first he told me that he intended to write a book that would cover economics, energy, intermediate technology and organisation, I doubted whether it could be effectively done. My scepticism was received with his usual cheerful confidence and *Small is Beautiful*[2] did have the cohesion and unity discernible throughout his work — and his life.

This volume contains about one-third of his statements on the themes of energy, public ownership and management made during his twenty years at the National Coal Board, and more occasional contributions in the seven years from his leaving the organisation until his death.

It puzzles even people who knew him well why he should have chosen to spend so long in the coal industry. In fact there were several reasons which were important to him. First, energy is indispensable to society — not a luxury: coal is one major means of providing it. Furthermore, his work involved him in long-term planning commitments. Nothing worthwhile could have been accomplished quickly. Then there was the attraction to a socialist of helping to make public ownership work. Public enterprise was preferable, offering choices of action and purpose, to private industry, where the pursuit of profit was the only objective. Finally, he strongly believed that the intellectual was indebted to the community which, in effect, enabled him to be what he was. There was fulfilment for him in contributing to society's wellbeing.

But for most of his life he questioned the orthodox and criticised official thinking. It was a speech about nuclear power that turned out to be Schumacher's most controversial public statement. In October 1967, at the height of the coal versus nuclear debate and while the Labour Government's Fuel Policy White Paper was being drafted, he gave the Des Vœux Memorial Lecture at the annual conference of the National Society for Clean Air. He quoted in *Small is Beautiful* those parts of his speech in which he raised doubts about the safety of nuclear reactors, drew attention to the problem of the radioactive wastes and argued for watchfulness on the effects of this new technology.

The Coal Board have always agreed, and still believe, that development of nuclear energy is necessary: the size and speed of the programme is what they have opposed. Much of the material for this speech consisted of quotations from no fewer than thirty-five eminent authorities, including such people as Galbraith, Einstein and Schweitzer, as well as official reports, mainly to the American Government. He ended with this moving statement of his beliefs: 'The continuation of scientific advance in the direction of ever-increasing violence, culminating in nuclear fission and moving on to nuclear fusion, is a prospect of terror threatening the abolition of man. Yet it is not written in the stars that this must be the direction. There is also a life-giving and life-enhancing possibility, the conscious exploration and cultivation of all relatively non-violent, harmonious, organic methods of co-operating with that enormous, wonderful, incomprehensible system of God-given nature, of which we are a part and which we certainly have not made ourselves.'

The response to his lecture was immediate and angry. Officialdom was outraged. He was accused of having spoken irresponsibly and of being guilty of special pleading. There was a public rebuke by the Minister of Power, Richard Marsh. No official complaint was made to the Chairman of the Coal Board, Lord Robens (himself a former Labour Minister), though, presumably because it was well known that he always stood by his officials.

The storm made Schumacher, briefly, lose his customary serenity. After all, as he saw it, he had simply been repeating the views of a number of reputable people. Within a few days he defended himself in a letter to *The Times* which was a fine example of his skill as a courteous, but firm, polemicist.

Despite the fact that English was not his mother tongue, Schumacher used it with clarity and elegance. The quality of his material is evident from reading it on the page and his presentation technique was a model for all professional communicators, yet it was a simple one. He could communicate effectively whether his audience consisted of academics, working men, professional people, or schoolchildren. He would put himself in the position of his audience. How much did they know about the subject? Did they care about it? Would it be all right to use technical expressions? This is not to say that he ever talked down to his audience.

In the early years of nationalisation the Coal Board used to hold week-long summer schools outside the university term, mainly in Oxford, though Cambridge was also used. About 500 people from the industry, accommodated in half-a-dozen or so colleges, would come from all the coalfields: they included mineworkers, under-officials, managers, engineers, scientists, administrators, marketing men, lawyers, accountants, and industrial relations staff. Much of the business was done in syndicates but there were also plenary sessions attended by everyone, sometimes addressed by the Minister responsible for the industry, sometimes the Board Chairman, but always by Schumacher. The purpose of the schools was to encourage a greater cohesion within the new organisation; to define the main problems of the day; to explain current policies and the reasons for them; to seek everybody's views on the big issues; and generally to develop a feeling of common purpose.

Quotations from some of his summer school speeches have been included in this volume: as I hope the selection shows, their range was wide, and given the nature of the audience, it is difficult to imagine how the opportunity could have been better used. Although his voice was by no means as powerful as that of an experienced politician, he refused to use a microphone despite the size of the hall (usually in the University Examination Schools). People had to make an effort to hear, but listen they did. Invariably, whether they were graduates or had left school at fourteen, they went away feeling they had learned something fresh and important. And some of the people who were present remember now the things he told them nearly thirty years ago.

He rarely used jokes as such but humour was inherent in his style. He would demonstrate the absurdity of an argument: his timing was superb and laughs were frequent. The scarcely discernible accent was an asset: listening to it was pleasurable. Added to all that was an immense skill for apt analogy and exposition. I was a course member at the Coal Board Staff College in 1958 when a questioner tried to test, by referring to the then recently-discovered Saharan oilfield, Schumacher's argument that energy was bound to become scarce and more expensive in real terms in the foreseeable future. With a twinkling eye he threw a few figures on the board and concluded: 'You see, if we could discover about 120 Saharas every year, we could just about look after the expected *growth* in demand.'

Fittingly for a man of his beliefs, Schumacher used non-violent methods of disputation: even when the questioning was hostile, he maintained his good-natured serenity. Always his response would be that of a man who recognised that the questioner sincerely believed in his argument; but with gentle firmness Schumacher would counter it. A rare example of his use of scorn occurred in his comments on a study done for the European Coal and Steel Community in 1962. His adversaries were professionals — fellow economists — and he resented the claims they were making for their professional skills. He used splendid invective to counter them. 'I think the time has come when it is necessary to be quite blunt about this sort of thing: these figures are not worth the paper they are written on. They are a case of spurious verisimilitude which borders on mendacity. The whole study, from beginning to end, proceeds from tendentious assumptions to foregone conclusions.' This review appears out of historical order at the beginning of a chapter because it is of such importance to the whole theme of planning and forecasting, giving an indication of the kind of problem the industry was being plagued with at the time. In some of the other chapters I have similarly placed a statement of overriding importance to the theme at the beginning; otherwise the extracts are presented in chronological order.

Each chapter has one main theme. Some speeches and articles dealt with more than one subject, so I have distributed their contents over two or more chapters. I hope cross references in the footnotes will help readers who want to trace extracts which had a common source. Some extracts are quite short, containing the kernel of what Schumacher had to say about energy in a wider-ranging article or speech; where he was developing a theme that is particularly relevant here, longer extracts are included. The source of each item is given above it, so that the reader may have in mind the date on which Schumacher was speaking or writing. This seems important in the context of Schumacher's references to time-scales, and it emphasises just how far-sighted he was.

It will be seen that Schumacher rarely needed to revise his views on major issues. In particular he was consistent about the folly of our becoming too dependent for energy supplies on the Middle East. The earliest warnings he gave, albeit to coal industry audiences, were in 1952: there followed frequent statements to the wider public, including one almost on the eve of the 1973 Arab-Israeli War. In *Small is Beautiful* (published in the year of the conflict, ironically), he recorded how such warnings were treated with derision and contempt.

Only the coal producers (and Schumacher's voice was the most prominent) took seriously the formation in 1960 of the Organisation of Petroleum Exporting Countries (OPEC) controlling between 80 and 90 per cent of the world's supplies. Its purpose was two-fold: to raise prices and to ensure that the member countries' reserves would last long enough to allow them to develop new economic structures. The official view was that the producing countries would not combine long enough to allow them to make any real difficulty for the industrialised nations. Lord Robens, who was appointed Chairman of the Board the following year and was there for the rest of Schumacher's time, has described in his book *Ten Year Stint*[3] how politicians and civil servants in Britain dismissed the new organisation from serious consideration. They were not alone in Western Europe or in the rest of the Western world.

In the event this proved to be just about the most devastating miscalculation ever made in economic policy: it continues to affect the standard of living of the populations of the importing countries, where economic growth is still hampered by the four-fold increase in oil prices that followed the Arab-Israeli conflict of 1973, and foreign policy, especially of the United States, is heavily influenced by the fear that supplies will be withheld. Internal disturbances in Iran starting in 1978, and the violent conflict with Iraq two years later, show that there are more sources than one for political instability in the Middle East.

It may seem extraordinary that Schumacher should, throughout the twenty years before the oil-producing countries used their power for political as well as economic ends, have warned so accurately what would happen. He himself thought it self-evident from the Organisation's formation: what is really surprising is that the politicians and civil servants should have been so complacent.

Now, with the proof of history, one can only speculate about the reasons why his arguments were ignored.

Schumacher identified one of them when, as early as 1952, he spoke about primary production (of foodstuffs and raw materials for export) being regarded as unfashionable and suggestive of colonial status, while most nations sought to develop their manufacturing or secondary industry. Related to this idea is the belief that what is new must be better. Nuclear power was new; coal was certainly not. Therefore the new energy source was to be preferred. Schumacher often pointed out that primary production cannot be readily expanded. The coal industry and agriculture were not very attractive subjects for swift technical change.

Lord Robens thinks there was another reason why the politicians did little to prevent the contraction of the coal industry between 1957 and 1973. As an extractive industry, mining employs a lot of people: in his opinion, Ministers regarded it as a pool of manpower which was needed for the more modern growth industries. As it turned out, such industries as have been expanded are mainly capital, and not labour, intensive. And that is among the reasons for the present high unemployment in the United Kingdom.

Yet another barrier to sensible planning of the energy industries is that, whereas a Government has a life of at the most five years, energy investment has a long lead time. Capital provided now out of the country's limited resources will not bring a return for eight or ten years in the case of a new mine or a big new power station.

Whatever the reasons, the results for Britain and the other developed countries were calamitous economically and politically. Because of the neglect of the coal industry the choice of substituting indigenous energy for imported oil when prices soared from 1974 up to the present day was not available. The North Sea was not yet a significant producer and there was insufficient coal to replace imported oil, for example, in the all-important electricity supply industry; nor was there the capacity to burn it.

Industrial activity in Britain has never regained the levels achieved before 1974.

Schumacher's belief in the need to preserve alternative choices had been ignored and the British people are still having to live with the consequences.

1
The Principles of Energy Conservation

Introduction

In coalmining it is not an act of conservation to close a colliery which still has access to reserves, with the intention of re-opening it later. Pressure of the strata quickly closes the underground roadways, water fills the workings and only in extremely rare cases have abandoned collieries been brought back into production. The heavy costs of keeping a mine in operational condition are insupportable without the output to carry them. So the productive capacity of the industry is diminished by each closure unless the pit is effectively exhausted. In mining, conservation means maintaining productive capacity: reserves in the ground are worth little without the access to work them.

Economists more orthodox than Schumacher tend to be suspicious of the whole conservation argument. Theories based on manufacturing processes have no validity when applied to non-renewable resources like oil, coal and natural gas. As Schumacher pointed out, no economic theory of primary production exists and conservation of energy sources is regarded as uneconomic, since it is opposed to the maximisation of profits now.

1

From 'Energy supplies − the need for conservation', *Energy International*, September 1964

There is no substitute for energy; the whole edifice of modern life is built upon it. Although energy can be bought and sold like any other commodity, it is not 'just another commodity', but the precondition of all commodities, a basic factor equally with air, water and earth.

Unlike air, water and earth, however, which are arranged in an ingenious and remarkable regenerative cycle, energy is largely non-renewable. Coal, oil and natural gas — 94 per cent of the world's use of primary energy — are taken from Nature's larder without possibility of renewal. Only 6 per cent of the total — the largest element of which is water power — is a renewable resource.

The world reserves of coal, oil and natural gas are undoubtedly very large, but the fact remains that they are a once-for-all endowment which cannot be increased but only diminished. The ultimate size of this endowment is unknown but there is an economic criterion of extraction costs which makes all estimates of 'ultimate reserves' well-nigh meaningless.

Matters of energy policy concern not merely the industries directly involved in the job of production and distribution but also governments and indeed the public at large. Here, in fact, is a field in which one cannot let things rip, where one has every reason to mistrust the wisdom of mere market forces; where there is need for foresight, good husbandry, and conscious conservation.

The idea of conservation, an object of the utmost suspicion to economists as a class, has a higher validity when applied to the four basic factors of human existence than when related to the production of goods and services out of factors offered by man's environment. In dealing with matters of fuel and energy the idea of conservation has to be kept clearly and tenaciously in mind.

Consider a feasibility study of the year 2000. How much energy and in what form will it be required by the world? Will there be a buyers' market (as now) [1964] or a sellers' market (as six years ago) [in 1958]?

Over the last 36 years there has been a very substantial increase in the per capita consumption of fuel. A great increase in the world population is also generally predicted. The per capita increase in energy consumption is most striking where there have been successful efforts at industrialisation.

Will there be more or less of this during the next 36 years? If there is going to be less 'economic development' between 1964 and 2000 than there has been between 1928 and 1964 this would imply a dismal failure and disappointment of the most cherished economic aspirations of our time. Extrapolation of the past trend in per capita consumption, reinforced by the accelerated growth in population, would take us to an energy requirement of approximately 18,000m. tons of coal equivalent in the year 2000, compared with about 4,800m. tons at present.[1] This would be rather less than a four-fold increase in the next 36 years, compared with rather more than a three-fold increase during the last 36 years.

Taking the forecasts of energy requirements published during the last few years by various countries, the assumed world total of 18,000m. tons of coal equivalent in the year 2000 appears to be a somewhat modest estimate which the underdeveloped countries might consider to err on the side of defeatism.

Nevertheless, a possible requirement of 18,000m. tons a year is enough to baffle the imagination.

Current world fuel supplies are roughly as shown in Table 1.1.1.

TABLE 1.1.1

	Million tons coal equivalent	%
Coal and lignite	2,100	44
Oil	1,700	35
Natural gas	700	15
Water, power, etc.	300	6
Total	4,800	100

In recent years, oil has been the most dynamic and expansionist major supplier. Are there any likely limits to the expansion of oil? A precise answer is impossible. The ultimate reserves are unknown; the proven reserves are an uncertain quantity, depending on future developments of the recovery rate; the political factors which will determine the exploitability and the cost of exploitation of known or yet to be discovered reserves, are virtually unpredictable. There is likely to be enough oil 'for many years to come', but the richest and cheapest reserves are located in some of the world's most unstable countries.

Faced with such uncertainty, it is tempting to abandon the quest for a long-term view and simply to hope for the best. But vital decisions have to be made now. Oil is cheap and plentiful at present; should indigenous fuel resources be abandoned like half the collieries of Western Europe which at present price and cost levels may seem to be uneconomic? Or should a policy of conservation be followed, which may require costly measures of protection and subsidisation? A rational answer depends largely upon an appraisal of the long-term supply prospects from other sources of primary energy, such as imported oil. But the oil industry cannot, in fact, give cast-iron assurances relating to long-term supply.

2

From 1966 and all that. Talk to National Coal Board Headquarters staff, London, December 6, 1950

I should estimate that in this country, had there been no replacement of coal by oil, total internal coal consumption in 1937 might have been 3 per cent higher than it was and thus 3 per cent higher than in 1907. No, it was not oil and it was not any other source of fuel that made it possible to double total output without increasing coal consumption; it was the improvement in the efficiency of coal utilisation, an improvement largely enforced by the fact that coal was becoming steadily and remorselessly dearer. I think we should remember this when we come to consider the future.

Each ton of coal, on the average, enables us to produce some £50–£60 worth of goods and services. Without it, productive forces twenty times as valuable as the forces required to produce a ton of coal are doomed to idleness. Let no one think that

4

the economic importance of £3 worth of coal is the same as that of £3 worth of any other commodity. There are only two basic items in the world economy — food and fuel. All the rest are secondary.

A rise of thermal efficiency from 20 to 30 per cent is a 50 per cent improvement; to obtain another 50 per cent improvement, thermal efficiency would have to increase from 30 to 45 per cent. Such an advance would be a very great achievement. Yet I do not doubt that a great deal can be done provided there is a motive for doing it. In the past, the improvement in fuel efficiency was driven forward by a steady and remorseless rise in the real price of coal. But in the last few years — contrary to what most people think — the real price of coal has fallen. In terms of other commodities, believe it or not, coal is now cheaper than it was in 1938. Is this a good thing when there is a shortage of coal, a shortage which makes it more than ever necessary that coal should be most efficiently and sparingly used?

3

From Long-term demands for fuel. Paper to a study group of the Royal Statistical Society, May 21, 1958[1]

I started by quoting the great Professor Jevons.[2] He asked the courageous question: 'Are we wise in allowing the commerce of this country to rise beyond the point at which it can be long maintained?' Instead of forecasting what we 'need', some people at least might attempt to work out what it would be wise and prudent to need, so that the way of life of this country 'can be long maintained'. How long? Yes, that is the question. Modern man seems to be singularly unwilling to face the future. Is it 'Après nous le fallout'? If one talks today of the year 2000, he is quite unwilling to listen, unless one is talking space travel or other bits of science fiction. Yet, as the late Sir Francis Simon[3] said four years ago, 'the twilight of the fuel gods will be upon us in the not very distant future'. It is precisely here that economists and statisticians could make a really vital contribution.

Instead of concocting long-term forecasts of requirements, they might make long-term studies of availabilities; they might help to open peoples' eyes to the fact that the problem of resources is in no way solved and that the way we are carrying on exposes our own children to totally insoluble problems. They might, furthermore, devote themselves to the study of alternative patterns of living — because unquestionably patterns exist which permit genuine increases in living standards without an increase in the squandering of irreplaceable resources.

4

From Coal — the next fifty years. Paper to a study conference organised by the National Union of Mineworkers, London, March 25, 1960

When applied to renewable goods such terms as 'cost of production', 'depreciation', 'economic or uneconomic' have a fairly clear meaning. For instance, it is 'uneconomic' to cultivate area 'A' if better results could be obtained by cultivating area 'B' instead and if only one of the two areas is needed to meet demand. Is the same true when 'A' and 'B' represent coal deposits, that is, non-renewable resources? Coal 'B' may indeed be cheaper to get than coal 'A' but it can only be got once. When it is gone, coal 'A' must be resorted to, whether we like it or not. In other words, we are not choosing to use 'B' *instead* of 'A', but merely to take 'B' *before* taking 'A'. This is not an economic choice at all. 'Best seams first' is not a principle of economics.* The master of the wedding feast who gives the best wine first and leaves the poor wines for later cannot invoke the principles of economics for his actions, and he who leaves the best wines for the end cannot be accused of acting uneconomically.

Once we recognise that the coal industry is not something like a manufacturing concern; that it is an extractive industry working a non-renewable asset of finite size *for which there is, as yet, no substitute in sight*, we find that the term 'uneconomic' may be highly misleading. An 'uneconomic' colliery may represent nothing more than a slight deviation from the principle of 'best

6

seams first'. This principle . . . is not a principle of economics. A deviation from the principle simply means that some of the better resources are being left for our children; that there is a certain amount of 'averaging' in the resources used by the successive generations.

'Picking the eyes out of the coal' has always offended the instinctive good sense and responsibility of mining engineers and mining communities. It cannot be right to 'sterilise' great masses of coal. Mining practices of the past which have led to such wasteful sterilisation are much criticised and regretted today. A mining industry that did not honour the concept of 'conservation' would be failing in its long-term duties to the community.

There is no need for this concept of conservation in the production of manufactured goods. The re-investment of depreciation allowances suffices to 'conserve' the production machine, and any thought of conservation beyond this ordinary practice is suspect. In particular, it is highly undesirable to 'conserve' any factory that fails consistently to earn a profit (unless there are extra-economic reasons for doing so). The idea of conservation becomes already broader and more urgent when we turn to agricultural production. A much higher degree of *responsibility for the long term* is needed here than in manufacturing industry where man is his own master. In agriculture, forestry, and fishing man must co-operate with nature, and great care has to be taken to conserve soil fertility, to safeguard forests, to maintain the population of certain fishes, etc. The dustbowl in America and the ravages of deforestation around the Mediterranean are telling witnesses of the insufficiency of the profit and loss account as the sole guide to action and of the inescapable need for the concept of conservation. But this concept achieves its greatest relevance and urgency when we consider non-renewable resources like coal in Western Europe. Here the highest degree of *responsibility for the long term* is needed in the public interest, as any waste of resources is an irretrievable loss of substance.

The idea of conservation was pungently stated by the Chairman of the National Coal Board[1] recently:

People ask whether I am still interested in fuel economy, in

7

the greatest possible efficiency in the use of fuels, now that there are large unsold stocks of coal on the ground. The answer is 'Yes'.

Anyone closely associated with the coal industry, anyone who knows what it takes to get the coal out of the ground, to get it prepared for the market and transported to the user, *anyone who knows coal as a wasting asset, as one of nature's irreplaceable bounties*, must hate to see coal wasted or used inefficiently.

If people think that, for commercial reasons, the coal industry should now ignore this and encourage extravagance, I disagree with them. (Italics are mine)

I think one conclusion is quite irresistible: that the continuance, beyond the year 2000, of the modern way of life in the most highly-developed countries of the West is virtually certain to depend on coal, and that it is therefore the inescapable duty of the present generation to ensure that the coal industry is cautiously and sensibly preserved in the meantime. This conclusion, for obvious reasons, applies with particular force, I suggest, to the United Kingdom.

I suggest that the only principle that is defensible on grounds of long-term expediency as well as morality would be the principle that (within reasonable limits) all existing collieries will be worked until the coal is gone. I am, however, bound to add at once that this principle, derived from considerations of *national* interest, cannot be implemented by the coal industry without *national* support. To put it differently: I suggest that what is needed is a clear recognition of the principle of conservation. This principle, however, is not always compatible with the most profitable operation in free competition; it therefore needs to be supported by *national* policy.

While it takes time to allow the industry to shrink (without wanton destruction), it takes even more time to expand it. To let the industry shrink now would indeed preserve resources for the 1980s; but unless the process of re-expansion is started in good time, it will not, in fact, prove possible to get more coal when it may be wanted. Not only does it take many years to

8

construct new collieries, there are also very difficult sociological factors to be taken into account. Where are the miners to come from in the 1980s when the mining communities have been destroyed − or let us say: dissolved − in the 1960s? I suggest that two assets are needed to get coal, a geological asset − coal reserves, and a sociological asset − a tradition of mining. If we want coal in the 1980s, we cannot afford to play havoc with these two assets now.

Please observe that it is a matter of assets, indeed: national assets. I personally believe that the proper preservation of these assets should be a matter of national policy. And that is the meaning which I should attach to the words 'national fuel policy'. Knowing that these assets will be needed in the not too distant future − and probably needed for survival − the nation should decide upon their preservation.

I suggest that it is a very different thing to preserve assets or to preserve merely an established activity. If an activity has become unrewarding, it will generally be wisest to stop it and find something else to do. The preservation of such an activity by means of national policy can seldom be defended. But the preservation of existing national assets *for which there is a future need* and which are irreplaceable − that surely is simply the exercise of ordinary foresight. It will cost a bit of money, unquestionably. It is not possible to provide for the future without drawing on the present. This country spends annually some £1,600m. on armaments, to preserve itself politically. Its economic survival demands the preservation of certain economic assets − the coal industry. All that is needed is a modest, reasonable retardation of the change-over to oil. The extra cost to British industry involved is so minute that it simply cannot be adduced as an argument.

The assets that require preservation are likely to be extremely valuable in twenty years or so. Even the most unprofitable colliery of today produces very cheap calories compared with the cost of calories likely to be obtained from any 'unconventional' process in twenty or thirty years' time. The national fuel policy that I have in mind would be animated by the conviction that Nature's deposits of concentrated carbon are a unique and irreplaceable bounty, a once-for-all endowment of mankind to ease its task of living, and that no generation has the right to

9

abandon and ruin any of them for the sake of a small, fleeting convenience.

5

From Some problems of coal economics. Paper to a seminar on problems of industrial economics, University of Birmingham, January 18, 1962[1]

The 'Economics of Primary Goods' are incomplete and misleading without the principle of conservation. Yet this principle is hardly ever encountered in textbooks on economics and this is no doubt due to the fact . . . that present day economics treats all goods as if they were manufactures. In manufacturing, to be sure, there is no need for any special idea of conservation. The reinvestment of depreciation allowances ought to suffice to 'conserve' the productive equipment, and any thought of conservation beyond this ordinary practice is highly suspect. In particular, it is most undesirable to 'conserve' any factory that fails to make a profit (unless it be for extra-economic reasons).

When we turn to primary production the idea of conservation immediately demands attention. Even with renewable primary products, man is not wholly in control of the productive process as he is in a factory but first must fit his actions into the rhythm of the seasons and the often mysterious requirements of organic life. His responsibility cannot be confined to making ends meet and maintaining his man-made assets: he must also 'conserve' the natural conditions which make primary production possible. Thus he has to conserve soil fertility, to conserve forests, to conserve fish populations, and so on. The need for conservation must take precedence over any requirements of the market if the real public interest is to be served.

The idea of conservation in the fullest sense applies to the 'getting' of non-renewable primary goods. Here the highest degree of responsibility is needed in the public interest as any waste of resources is an irretrievable loss of substance. This means that production policy must be largely insulated from the impact of market forces, for the market is concerned only with the short-term 'bargain' aspect of things and not with the underlying reality.

10

This problem is particularly acute in the coal industry for both geological and sociological reasons. It is not possible to start and stop coal-getting operations at short notice to meet short-term market fluctuations. The attempt to do so invariably causes long-term damage and a spoiling of reserves. Proper conservation requires that operations should be planned carefully and steadily over long periods of time and that all sudden rushes of expansion or contraction should be avoided.

The criteria of efficiency in coal production, therefore, cannot be the same as those applicable to secondary industry where there is no problem of conservation. I am looking forward to the day when academic economists will work out a proper 'Theory of Goods' which distinguishes and recognises at least the special features of non-renewable primary goods. At present no such theory exists; economic policies are judged as if it made no difference whether the output of an enterprise was coal, butter, nylon stockings or haircuts. The coal industry is criticised for every one of its policies of conservation, for every show of a responsible husbandry of its irreplaceable resources, for accumulating undistributed stocks, for refusing to rush into wholesale closures, for trying to preserve certain mining communities — in short for every policy designed to ensure that Britain's coal reserves and her ability to exploit them will not be jeopardised by the impact of temporary and fortuitous market factors.

Coal's most serious competitors are oil and natural gas, both of them non-renewable resources. Hydro-electricity is the only renewable fuel of commercial importance; it accounts for hardly 7 per cent of the world's total fuel use. The future potential of atomic energy is still unknown but there is now a consensus of opinion that its contribution to the world's fuel requirements will be negligible for many years to come. Meanwhile, these requirements are growing steadily at compound interest rates. The annual average rate of growth over the last forty years was about 2 per cent cumulative, leading to a trebling of the world's fuel requirements in forty years. With world-wide industrialisation and urbanisation, supported by development on a large scale, it is difficult to imagine that the growth rate during the next forty years will be any smaller than it was in the last; on the contrary it might well be higher. In

11

other words, all established trends point to a fourfold increase in world fuel requirements by the year 2000, or shortly thereafter — that is, an increase from the present 4,500m. tons of coal equivalent to an annual requirement of about 18,000m. tons.

The failure of economic theory to distinguish non-renewable from renewable products may be one of the causes of the total unconcern with which such prospects are being treated. It is perfectly obvious that the whole of Western Europe — already the largest oil importer in the world — is in an increasingly vulnerable position as the world's fuel requirements continue to grow, and that the position would quickly become desperate if its coal industries were substantially curtailed. The simplest arithmetic shows that a fourfold increase in world fuel use, if possible at all, could be attained only by making the maximum use of all fuels, including coal. Oil and gas would quickly become premium fuels whose high price would limit their use to purposes which coal cannot serve, such as forms of transportation. Atomic energy would, in any case, have to be developed to the uttermost so that at least in the more advanced countries the base load could be provided thereby. Yet light-hearted and destructive talk about the coal industry being uneconomic goes on, suggesting that the industry's policies of conservation are nothing but the narrow-minded defence of a vested interest.

Once it has been accepted that conservation must be an important element of policy and that departures from the principle of 'best seams first' are not to be taken as sins against economic rationality, it is also easy to see that the choice of investment projects must be mainly a technical one. A misdirection of capital investment can be due to a technical or a geological error; if it is misdirection only in the sense that the second best, rather than the best, coal reserves are being tackled, no real harm is done; there is merely a slight shift of advantage from the present to the future, such as is involved in every decision to save and invest, rather than to consume. It is surprising that economists, who invariably applaud the latter, with the same invariability deplore the former.

It is obvious that conservation costs money in the short run, while its salutary effects in the long run are likely to pass unnoticed. Can the coal industry afford conservation at the present time when the pressure of oil competition is particularly severe? This question, however plausible it sounds, is based on wrong thinking. Cheap oil is not an emergency like war which could force the country into policies to maximise immediate advantage even at the risk of jeopardising the more distant future. Cheap oil is a benefit to the national economy and thus makes it easier to afford conservation.

But it is the economy as a whole that must bear the cost of it, just as the economy as a whole bears the cost of education, defence, farming subsidies and so on. If the coal industry is left without support and forced into catch-as-catch-can commercialism, it cannot protect the longer-term public interest through conservation policies but will have to pick the eyes out of the coal just as many private owners, under the pressure of competition, used to do in the nineteenth century.

6

From 'European coal in the year 2000', *Schlägel und Eisen,* February 1963 – translation

Energy demand will grow and not diminish in the future; it will have to be filled mainly by coal and oil; the reserves, because of the increased outputs, especially of oil and gas, are continually diminishing (even if reserves from new discoveries are increasing). The dependence of Western Europe on imports of these fuels from an increasingly energy-hungry world is steadily rising.

The dominant question therefore is: on whom will the West European economy be dependent? So long as we do not wish to depend on fuel imports from the Soviet bloc, they must come from the oil-rich states that have recently joined together to form the Organisation of Petroleum Exporting Countries, OPEC, which controls 85 to 90 per cent of the world's oil exports.

The strength of OPEC lies, above all, in the logic of their reasoning. 'Our oil is all we have', they say. 'If it is used up in a few decades, we shall be lost. We must therefore ensure that it lasts long enough to allow us to develop another way of life.' This means less oil and at higher prices. Lowering of prices, often dreamt of because of the extraordinarily low production costs in the Middle East, is not likely both for this reason and because of the steadily rising costs of new discoveries. For apart from the Soviet bloc and the Middle East, only in Canada and the Sahara have there been new oil finds in the last few years and the quantities discovered there would cover world consumption only for about 18 months. These scanty results almost all over the world at a time of the most intense drilling activity to depths never previously reached, arouse fears that with much faster growth of consumption the world's oil supply will not even be ensured for the next twenty years, certainly not at current prices.

According to all the conclusions of a careful investigation of the future, one must accept that the world is approaching a period of energy shortage that will be tied to lasting price increases for all imported fuels. This will not occur immediately but only after some years. In the most favourable conditions of a peaceful, successful, rapidly-growing world economy, we would have to count on an acute shortage of fuels within twenty years. If this should be a true diagnosis, the need for a careful, responsible policy of conservation of the European coal industry should no longer be in doubt. If Europe needs to rely on a coal output of 400 to 500m. tonnes in 1980, she must be ready to maintain a capacity approaching this level in the 1960s and 1970s. All current colliery closures must be significant, not just in the light of the 1962 home energy situation, but also for the world energy situation expected in the years 1975–85. It is clear that not only the position in the world, but also the very existence of Western Europe would be seriously endangered if she were to be faced with an unfillable energy gap. Such a situation does not need to last long before it shakes political and economic institutions to their foundations.

Considering this unique danger to the whole West European economy, that is to the toil, hopes, and working aspirations of 300m. people living close together, one can only wonder at the frivolity with which people talk sometimes about 'the long-term security of West European energy supplies'. This security is said to come not through the safekeeping of what is in one's own country (its indigenous fuel resources), but through the so-called 'de-centralisation of supplies' and a 'stocking policy for oil imports'. As for the much-acclaimed diversification of supplies one can only say that, apart from the opening of the Saharan wells, it consists merely of substituting Middle East for Soviet oil, and with the approaching world shortage, even the most far-flung diversification would be ineffectual. So far as a 'stocking policy for oil imports' is concerned, a serious investigation of the order of magnitude is recommended. In 1970 West European fuel imports, if the coal mining industry is maintained, will amount to about 650m. tonnes of coal equivalent. Should therefore a six months' supply of oil supply be stocked, i.e., 300m. tonnes of oil? The area will need both the securities of diversification of imports and a stocking policy even if it maintains its coal industry to the full extent. These 'additional insurances' are of limited use and can in no way replace the security of home production. The security of the energy supply of the West European economy is of vital interest to all who live there. What is at stake here far surpasses any conflict of interest between coal and oil.

Western Europe's hopes of maintaining its prosperity and lifestyle to the end of the century and beyond must rest on a permanently adequate, accessible fuel supply, that is, in the first place, on the West European coal industry. This must be conserved and used with care.

7

From 'The need for long-term thinking in fuel and power', *Revue française de l'énergie,* April/May 1964[1]

There can be no real doubt about the kind of energy policy that it would be rational for Western Europe to adopt. It would in its essence be a policy of careful and judicious *conservation* of the indigenous primary fuel industries.

The first requirement of a conservation policy with regard to an extractive industry is the avoidance of rapid change. Only when working to a steady programme can wasteful extraction

15

and the ruin of irreplaceable resources be avoided. The layman may think that the best way to 'conserve' indigenous coal resources is not to work them, taking imported fuel instead. This is true where work has not already started: an undisturbed coal deposit, of course, will conserve itself. But it is not true where collieries have already been developed and their production is stopped for the lack of markets. The closure of such collieries in the vast majority of cases means their abandonment virtually for all time. With the closing of the colliery its coal reserves normally are sterilised, that is to say, they become unworkable, except at costs which even in a world of much higher fuel prices would be unbearable.

As total fuel and energy requirements are rapidly growing, a policy of conservation is not at all difficult to carry out. The fastest growing sector is electricity generation, and there can be little doubt that the time will come when Western Europe's indigenous fuel supplies will be unable to cope with the growing requirements of the power stations, so that there will be no alternative to using imported fuel. All that a conservation policy requires is that the indigenous coal industries should be given preference over overseas fuel suppliers until their 'order books' are full. In other words, it is simply a matter of proper timing: take the indigenous product first and the imported product later, because, if you take the imported product now the indigenous industry will die and there will be no indigenous product later.

The simple truth of the matter is that all fuels will be needed. The problem is not how to cope with a long-term surplus, but how to avert a shortage. The trouble is that the market situation of the last few years in no way reflects the longer-term realities; it is nothing more than the temporary effect of the American oil import restrictions and of the internecine war of the oil companies among themselves.

It is precisely in such a situation that one would hope for governments to intervene for the purpose of safeguarding the position in the long run. It is not a question of protecting any particular vested interest; fuel and power is everybody's vital interest. There is a consensus of informed opinion that all primary fuel resources which Nature or science can make available will be needed. What could be more rational than a policy which ensures that none will be prematurely abandoned?

16

8

From Clean air and future energy — economics and conservation. Des Vœux Memorial Lecture to National Society for Clean Air, Blackpool, October 19, 1967[1]

It is hardly an exaggeration to say that, with increasing affluence, economics has moved into the very centre of public concern, and economic performance, economic growth, economic expansion, and so forth have become the abiding interest, if not the obsession of all modern societies. In the current vocabulary of condemnation there are few words as final and conclusive as the word 'uneconomic'. If an activity has been branded as uneconomic, its right to existence is not merely questioned but energetically denied. Anything that is found to be an impediment to economic growth is a shameful thing, and if people cling to it, they are thought of as either saboteurs or fools. Call a thing immoral or ugly, soul-destroying or a degradation of man, a peril to the peace of the world or to the well-being of future generations; as long as you have not shown it to be 'uneconomic' you have not really questioned its right to exist, grow and prosper.

But what does it mean when we say something is uneconomic? I am not asking what most people mean when they say this; because that is clear enough. They simply mean that it is like an illness: you are better off without it. The economist is supposed to be able to diagnose the illness and then, with luck and skill, remove it. Admittedly, economists often disagree among each other about the diagnosis and, even more frequently, about the cure; but that merely proves that the subject matter is uncommonly difficult and economists, like other humans, are fallible.

No, I am asking what it means, what sort of meaning the method of economics actually produces. And the answer to this question cannot be in doubt: something is uneconomic when it fails to earn an adequate profit in terms of money. The method of economics does not, and cannot, produce any other meaning. Numerous attempts have been made to obscure this fact, and they have caused a very great deal of confusion; but the fact

17

remains. Society, or a group or individual within society, may decide to hang on to an activity or asset for non-economic reasons — social, aesthetic, moral, or political — but this does in no way alter their uneconomic character. The judgment of economics, in other words, is an extremely fragmentary judgment; out of the large number of aspects which in real life have to be seen and judged together before a decision can be taken, economics supplies only one — whether a thing yields a money profit to those who undertake it or not.

Do not overlook the words 'to those who undertake it'. It is a great error to assume, for instance, that the methodology of economics is normally applied to determine whether an activity carried on by a group within society yields a profit to society as a whole. Even nationalised industries are not considered from this more comprehensive point of view. Every one of them is given a financial target — which is, in fact, an obligation* — and is expected to pursue this target without regard to any damage it might be inflicting on other parts of the economy. In fact, the prevailing creed, held with equal fervour by all political parties, is that the common good will necessarily be maximised if everybody, every industry and trade, whether nationalised or not, strives to earn an acceptable 'return' on the capital employed. Not even Adam Smith had a more implicit faith in the 'hidden hand' to ensure that 'what is good for General Motors is good for the United States'.

However that may be, about the fragmentary nature of the judgments of economics there can be no doubt whatever. Even within the narrow compass of the economic calculus, these judgments are necessarily and methodically narrow. For one thing, they give vastly more weight to the short than to the long term, because in the long term, as Keynes put it with cheerful brutality, we are all dead. And then, secondly, they are based on a definition of cost which excludes all 'free goods', that is to say, the entire God-given environment, except for those parts of it that have been privately appropriated. This means that an activity can be economic although it plays hell with the environment, and that a competing activity, if at some cost it protects and conserves the environment, will be uneconomic.

It is obvious that the idea of conservation is more than ever in

18

need of support, as the tempestuous advances of science and technology multiply the hazards. But as I said before, it is an uneconomic idea and has therefore no acknowledged place in a society under the dictatorship of economics. When it is occasionally introduced into the discussion, it tends to be treated not merely as a stranger but as an undesirable alien, probably dishonest and almost certainly immoral. In the past, when religion taught men to look upon Nature as God's handiwork, the idea of conservation was too self-evident to require special emphasis. But now that the religion of economics lends respectability to man's inborn envy and greed and Nature is looked upon as man's quarry to be used and abused without let or hindrance, what could be more important than an explicit theory of conservation? We teach our children that science and technology are the instruments for man's battle with Nature, but forget to warn them that, being himself a part of Nature, man could easily be on the losing side.

Modern economic thinking, as I have said, is peculiarly unable to consider the long term and to appreciate man's dependence on the natural world. It is therefore peculiarly defenceless against forces which produce a gradual and cumulative deterioration in the environment. Take the phenomenon of urbanisation. It can be assumed that no one moves from the countryside into the city unless he expects to gain a more or less immediate personal advantage therefrom. His move, therefore, is economic, and any measure to inhibit the move would be uneconomic. In particular, to make it worthwhile for him to stay in agriculture by means of tariffs or subsidies, would be grossly uneconomic. That it is done none the less is attributed to the irrationality of political pressures. But what about the irrationality of cities with millions of inhabitants? What about the cost, frustration, congestion and ill-health of the modern monster city? Yes, indeed, these are problems to be looked at but (we are told) they do not invalidate the doctrine that subsidised farming is grossly uneconomic.

It is not surprising, therefore, that all around us the most appalling malpractices and malformations are growing up, the growth of which is not being inhibited, because to do so would be uneconomic. Something like an explosion has to occur before warning voices are listened to, the voices of people who

19

had been ridiculed for years and years as nostalgic, reactionary, unpractical and starry-eyed. No one would apply these epithets today to those who for so many years had raised their voices against the heedless economism which has turned all large American cities into seedbeds of riots and civil war. Now that it is almost too late, popular comments are outspoken enough: 'Throughout the US, the big cities are scarred by slums, hobbled by inadequate mass transportation, starved for sufficient finances, torn by racial strife, half-choked by polluted air.' And yet: 'The nation's urban population is expected to double by the beginning of the next century.'† You might be tempted to ask, Why? The answer would come back: Because it would be uneconomic to attempt to resettle the rural areas. The American economist, John Kenneth Galbraith, has brilliantly shown how the conventional wisdom of economics produces the absurdity of 'private opulence and public squalor'.‡

Other changes, equally destructive or even more so, are going on all around us, but they must not be talked about because to do so might cause alarm and even impede economic growth. All the same, we cannot claim that we have not been warned. For instance, in spite of enormous advances in medicine, on which we do not fail to congratulate ourselves, there is a relentless advance in the frequency of chronic illness. The US Public Health Service states that, 'About 40.9 per cent of persons living in the United States were reported to have one or more chronic conditions. While some of these conditions were relatively minor, others were serious conditions such as heart disease, diabetes, or mental illness.'§

It is not my purpose to investigate the causes of this extraordinary development. It is well known that the infectious diseases, which were the principal causes of death in 1900, have been reduced almost to vanishing point; but that deaths from the so-called degenerative diseases have greatly increased, particularly deaths from cancer, heart disease, and diabetes, involving increasing numbers of children and young adults.

Developments of this kind are invariably the result of imbalance and disharmony. In the blind pursuit of immediate monetary gains modern man has not only divorced himself from nature by an excessive and hurtful degree of urbanisation,

20

he has also abandoned the idea of living in harmony with the myriad forms of plant and animal life on which his own survival depends; he has developed chemical substances which are unknown to nature and do not fit into her immensely complex system of checks and balances; many of them are extremely toxic, but he none the less applies them or discharges them into the environment, as if they would be out of action when they had fulfilled their specific purpose or could no longer be seen.

The religion of economics, at the same time, promotes an idolatry of rapid change, unaffected by the elementary truism that a change which is not an unquestionable improvement is a doubtful blessing. The burden of proof is placed on those who take the 'ecological viewpoint': unless they can produce evidence of marked injury to man, the change will proceed. Common sense, on the contrary, would suggest that the burden of proof should lie on the man who wants to introduce a change; he has to demonstrate that there cannot be any damaging consequences. But this would take too much time, and would therefore be uneconomic. Ecology, indeed, ought to be a compulsory subject for all economists, whether professionals or laymen, as this might serve to restore at least a modicum of balance. For ecology holds

> that an environmental setting developed over millions of years must be considered to have some merit. Anything so complicated as a planet inhabited by more than a million and a half species of plants and animals, all of them living together in a more or less balanced equilibrium in which they continuously use and re-use the same molecules of the soil and air, cannot be improved by aimless and uninformed tinkering. All changes in a complex mechanism involve some risk and should be undertaken only after careful study of all the facts available. Changes should be made on a small scale first so as to provide a test before they are widely applied. When information is incomplete, changes should stay close to the natural processes which have in their favour the indisputable evidence of having supported life for a very long time.**

21

Letter to *The Times* published on October 25, 1967

There has been so much shocked comment on my lecture to the National Society for Clean Air last week[1] that you may perhaps be good enough to allow me space for two explanations. First of all, nothing was further from my mind than to suggest that the Ministry of Power or the CEGB[2] or the Atomic Energy Authority had disregarded safety considerations in the development of nuclear power or that there existed a 'conspiracy of silence' regarding them. Frankly, I do not see how the words I used could bear any such interpretation, although I accept, of course, that the misunderstandings that have occurred must be due to faults in my exposition.

Secondly, what I did mean to suggest was something quite different. Processes involving ionising radiation are generally admitted to be inherently dangerous in a very special way in that they are capable of causing genetic damage to all living beings including humans. Assuming now that we can freely choose between two processes for doing a certain job, process A involving ionising radiation on a very large scale — and therefore requiring novel safety measures of the most sophisticated and exacting kind — and process B involving physical or chemical reactions of a relatively harmless kind with which we have long learned to cope more or less adequately: should we allow our choice between A and B to be exclusively or even preponderantly determined by comparative cheapness?

I ventured to suggest that economic considerations must not automatically be accepted as decisive in such a case. I suggested, in particular, that there was no need to be in a hurry to change over from conventional power stations to nuclear stations, that is to say, from processes with which we have learned to live, to a process capable of damaging the genetic properties of the human race and disturbing the environment in incalculable ways. The only excuse for haste in such a matter could be an inescapable necessity — which obviously does not exist. So, then, why not take much more time over research and experimentation before moving into the massive application of such an inherently dangerous process?

I am not alone in taking the view that — in the absence of necessity — even a small amount of genetic damage cannot be justified or excused by economic considerations. That this view, if accepted, would greatly mitigate the almost desperate problems of the coal industry and could also save the country a great deal of money, does not make it either illegitimate or absurd.

10

From a film produced by the National Coal Board, 1972, about district heating[1]

Fuel — most of it is oil, gas and coal — has only a limited quantity. We may not know precisely how much it is in the world, but it's limited, whereas our needs, the needs of our children, are unlimited. And therefore, I think waste is a central concept the moment we talk about fuel, and anything that eliminates waste must be very, very seriously considered. I believe that fuel is going to get very much dearer; of course we see it now, it has been rising very, very rapidly. Not only coal, but oil, American coal, nuclear energy, gas I'm not so sure about. But the notion which has been maintained for the last twelve years, that we are entering into an era of cheap fuel, that's a lot of nonsense. We're entering into an era of dear fuel. This is an important point, because in district heating, most of the cost is the pipes, the installation, the capital cost, and the running costs in terms of fuel is relatively small. So if fuel gets much dearer, and the capital cost doesn't rise very much, district heating will have an advantage over all other types of heating.

11

From 'Western Europe's energy crisis — a problem of life-styles', *Ambio,* Royal Swedish Academy of Sciences, 1973[1]

Needless to say, steeply-rising oil prices will, to some extent, curtail demand, but, as already mentioned, fuel demand is highly inelastic, because fuel is primarily a means of production for which there is no substitute and no possibility of re-cycling, so that going without fuel means going without production. The most expensive fuel, one might say, is the fuel one cannot get when one needs it. There are, of course, many possibilities of improving the technical efficiency of fuel and energy utilisation, which during the era of cheap fossil fuel supplies have been somewhat neglected. But it would be fanciful to attribute to these possibilities a decisive quantitative significance: their realisation requires very large capital investments — as for instance in the utilisation of waste heat from existing power stations — and, what is more, a great deal of time. There is no realistic possibility for improvements in fuel efficiency to do more than slightly reduce the rate of increase in total fuel and energy demand. To try and do so is, of course, helpful; but it cannot be expected to make more than a marginal contribution.

The development policies of the last twenty years have been virtually exclusively based on the assumption that 'development' can be most speedily achieved by transferring the high technology of the rich countries to the Third World. Where this transfer has been effected, the result has been a concentration of development upon big cities; a massive migration of rural populations into these cities which consequently have become infested with enormous slums; mass unemployment; stagnation of life in the rural areas; and sharply increasing energy requirements. The view is now gaining ground that what the Third World needs more urgently than anything else is an 'appropriate technology', although there is as yet little understanding as to what constitutes 'appropriateness'. What emerges ever more clearly from the work of the London-based Intermediate Technology Development Group, established in 1965, is that one of the primary criteria of 'appropriateness' is 'small-scale'. Most people of the Third World still live in villages and small towns, and they cannot possibly be absorbed into cities. Local markets are generally small, both on account of decentralised living and on account of poverty, which means that large-scale mass-production industry cannot economically be fitted into the rural areas. Small markets demand small-scale

24

production units, and these can be viable only on the basis of an efficient small-scale technology. Practical work on the development of such technologies has already established their feasibility, provided that the best resources of modern science and technical knowledge are deployed to this end.

12

From 'Energy and poverty'. *Coal and Energy Quarterly,* winter, 1974

The modern world is running out of cheap energy and may be also running out of time to do something about it. This will have far-reaching consequences, for a century of cheap and plentiful energy supplies has left us with modes of production, standards of consumption, and patterns of human settlement which are in general ill adapted to conditions of high energy costs and/or energy scarcity.

When essential and indispensable supplies rise in cost, it is invariably the poor who suffer the most, in spite of the fact that they use the least. This brutal truism is illustrated by Table 1.12.1, which was compiled by the Washington Center for Metropolitan Studies.*

TABLE 1.12.1 *Percentage of family income spent on energy declines as income increases*

Income status	Median household income ($)	BTU's per household	Cost per household ($)	% of total income
Poor	1,950	207	379	19·89
Lower-middle	7,960	294	572	8·09
Upper-middle	14,000	403	832	5·95
Well-off	22,000	478	994	4·42

The impact of, say, a trebling of energy costs on people who already have to devote 19·89 per cent of their total income to this item is, of course, infinitely more severe than on people who spend only 4·42 per cent in this way. What applies to households applies also to nations, *provided* the life styles of nations show the same similarity among themselves as the life styles of households.

The devastating effect of increased energy costs on poor countries can already be observed in agriculture. The 'Green Revolution', based on the successful development of marvellously productive new seeds, has been the world's main hope of expanding food output at least in line with expanding world populations. Now, however, in many parts of the world the 'Green Revolution' has virtually collapsed as a result of shortages of fertilisers and of power to drive the irrigation pumps. Traditional agriculture, of course, is relatively immune to changes on the energy front as it is virtually independent of vital inputs from large industrial systems. But traditional agriculture tends to be insufficiently productive to meet the growing food requirements of rapidly-increasing populations. Modern 'Green Revolution' agriculture is highly productive, but it suffers from at least three serious disadvantages: it is 'accessible' only to farmers who are already in a favourable position on account of fertile soil, ample water supplies, and considerable wealth; it does little for the longer-term maintenance of soil fertility; and it is totally dependent on cheap and plentiful energy supplies. Here, as in so many other cases, the *modern* mode of production benefited only small minorities already favourably placed but did nothing for the great masses of the poor whose situation became worse not merely in relative but also in absolute terms.

It can fairly easily be calculated that the modern system of agricultural technology is so greatly dependent on fossil fuels, primarily oil and natural gas, that the attempt to feed the whole of mankind — say, 4,000m. people, not allowing for further increases in world population — by means of these technologies, would lead to the absorption of all known oil and natural gas reserves *by agriculture alone* within a few decades. The only hope of survival for the poor in many parts of the world lies in the *upgrading* of traditional farming methods, with a strict and intelligent observance of the 'law of return', a meticulous conservation of water, and the fullest use of 'the human factor' rather than resort to large-scale mechanisation. In most countries, this inescapable truth has not yet been learned; and it is to be expected that the cost of this failure will be widespread famine conditions in the very near future.

Among the 'structures' produced by and inherited from a century of cheap and plentiful fossil fuels, the most striking and certainly the most visible are enormous cities and conurbations with many hundreds of thousands and even many millions of inhabitants. Although man has been building cities for 5,000 to 6,000 years, the monster cities which we now accept as normal are a good deal less than 100 years old, and all too many of them are to be found in poor countries.

Professor Kingsley Davis, one of the best-known students of the subject, said a few years ago: 'Clearly the world as a whole is not fully urbanised, but it soon will be.' This prediction, I suspect, is based on little more than an extrapolation of a 100-year trend. But it is not permissible to extrapolate trends unless one can be reasonably certain that the *material* factors which produced the trend and have kept it going are still operative. It is easy to see that cities live from the products of a 'hinterland' of rural areas; the bigger the city, the bigger has to be the hinterland from which it draws its supplies; and the bigger the hinterland, the greater is the city's dependence on fast, long-distance transport. Now, transport today is almost totally dependent on fossil fuels, primarily oil. A marked rise in oil prices — not to mention possible oil scarcity — fundamentally alters the economics of city life, not only on account of the transport factor just mentioned but also because many of the city buildings — high-rise flats, hotels, office blocks, and so forth, are themselves, as it were, huge *engines* dependent for their functioning on a continuous high rate of energy consumption.

If cheapness and plenty, as regards fuel, have produced these expensive structures, what is going to be the effect of dearness and scarcity, particularly in the poor societies of the Third World? *High density living patterns can be sustained only by high density fuels*, unless each unit is kept to a very modest scale. If the availability of the latter (for example, oil) is called into question, the viability of the former is called into question. If high density fuels become very expensive and hard to obtain, high density living in large cities becomes increasingly insupportable. Here again, it is virtually inevitable that the poor will be the main sufferers.

The change in the fuel situation from cheapness and plenty to dearness and scarcity calls for new types of calculation and new criteria of efficiency. The technology that has come into being during the period of cheapness and plenty is unlikely to fit the new conditions. Needless to say, everything calculable in terms of money will have to continue being calculated in money; but most of these money calculations will have to be supplemented and checked against calculations made in fuel units, such as calories. Poor societies — and probably the rich as well — may find that they are not able to afford things which are expensive in terms of calories, even if they appear to be relatively cheap in terms of money. There is not only a money cost of fuel but also a *fuel cost of fuel*. The Bureau of Mines in the United States recently began a study to determine how many calories it takes to produce the calories of various types of fuel: what matters is not *gross* energy production but net *energy gain*. Professor Odum of the University of Florida claims that 'the biggest lesson to be learned from net energy thinking is that all the new technologies being developed to attain energy independence are draining present energy supplies and are therefore hastening the day when fossil fuels run out. For example, enriching uranium for light-water reactors consumes, in the form of coal, 60 per cent of the energy released from the nuclear fuel'. Whether any nuclear energy producer is, in fact, a *net energy producer* at all, is still very much an open question.

A century of cheap and plentiful fossil fuel has left us not only with a great variety of structures and life styles, many of which are ill adapted to the conditions of dearness and scarcity of energy which are now emerging: it has also left us with an enormous body of scientific knowledge and technical experience, including a uniquely powerful methodology of problem solving. The hope for the future — certainly as far as the poor countries are concerned — lies exclusively in the existence of these assets, but it can be realised only if our problem-solving abilities are engaged *on the right problems*.

More and more of the poor countries are beginning to realise that the highly sophisticated, immensely capital-intensive mass-production technology of the rich societies, *which has been developed during the century of cheap and plentiful fossil fuels*, is extremely ill-suited for the development of poor societies. It *presupposes*, among other things, the existence of a highly elaborate infrastructure, of a ready availability of huge amounts of capital, and of large, wealthy, highly organised markets. None of these presuppositions can be fulfilled in poor countries, except possibly in a few big cities. The rich man's technology tends to 'eliminate the human factor from the productive process', which is handed over to machines totally dependent on cheap and plentiful fuel, while mass production entails mass transport, also totally dependent on cheap and plentiful fuel.

More and more of the poor countries are therefore asking for help in the development of alternative technologies, of modes of production and patterns of consumption which enable poor communities *to help themselves* by means, primarily, of local production from local materials for local use. Considerable progress in the development of such alternative technologies has already been made — for instance by the Intermediate Technology Development Group, established for this purpose ten years ago in London.

The work of the Group demonstrates the possibility of achieving efficient production by small-scale units of rigorously simplified design, low capital requirements, and easy maintenance. Production and consumption are thereby brought closer together; recycling becomes relatively easy; transport requirements are radically reduced; and the greatest productive force, plentifully available in poor societies, the 'human factor', can be re-integrated into the productive process. In addition, *small energy sources*, such as sun, wind, 'bio-gas' (methane), small amounts of flowing water, etc., can be harnessed to the benefit of man.

It appears that the only possibility of meeting the awesome challenges presented by the change in the world energy situation lies in the pursuit of such alternatives. This is unquestionably true of the world's poor societies, and it may yet turn out to be true even for the rich.

2

The World's Energy Needs and Resources

Introduction

It is now a decade since Schumacher left the National Coal Board and ceased to be a regular commentator on energy matters. Has anything happened since then to invalidate his warnings of ultimate crisis based on the failure of proved reserves of oil to keep pace with the increase in its use? He had little to say about North Sea oil which started to come ashore in quantity only five years ago. Its successful development has made Britain self-sufficient in energy supplies — a condition envied by less fortunate industrialised countries. Western Europe, which takes nearly half the output, has gained a stable, if not cheap, new source.

Would Schumacher have moderated his warnings if he had known this would happen?

It is against the insatiable world demand for oil that the contribution of the North Sea has to be assessed. Published figures for its proven reserves are sufficient to support an output at the present rate of 80m. tonnes a year until 1985, after which production will start to fall unless more big fields are proved. If all probable and possible reserves turn out to be recoverable (which would be astonishing), and full economic advantage is taken of them so that output is doubled for a time, it would still start to fall off before 1990. Given a policy of conservation, with output being increased only to 150m. tonnes a year, again based on the most optimistic estimates of reserves, it would still start to drop quickly before the year 2000.

In fact, the North Sea reserves, economically essential though they are to Britain, represent an addition of only 4 per cent to the world total. There has been no new major oilfield proved anywhere since 1973. In 1977, huge probable reserves in Mexico were announced: even if they all eventually prove recoverable, the world's stock of oil will rise by only 10 per cent.

So the Schumacher analysis of certain crisis remains robust; nothing that he said on the subject of energy needs and resources requires any qualification in 1980.

There is now a feverish effort to find processes for making liquid fuels from coal. At the last count, there were no fewer than 110 alternative fuel projects, most of them in the United States, and most involving oil companies or their coal subsidiaries. The US Department of Energy is aiming at an annual output of 100m. tonnes of synthetic fuels by 1982 (against a present use of 86m. tonnes a year of oil products, about half of which has to be imported!).[1] The European Economic Community is putting about £65m. into projects for coal liquefaction and gasification; West Germany is making a huge effort estimated to cost £3,000m.; Great Britain has two promising NCB processes for which pilot plant design studies have been completed; and even Sweden, with no coal resources, is spending money in the same quest for alternatives to oil. There is a hectic race to be the first with marketable technology.

Schumacher would have smiled wryly at all this: as long ago as 1954 he foresaw that before the end of the century there would be a need for liquid fuels synthesised from coal. So in this respect, his thinking has become part of the conventional wisdom.

In the same 1954 paper, he wrote of the possibility of Japan gaining access to Chinese coal: at the present time that resourceful country is seeking to overcome the political difficulties of investing simultaneously in coal and other energy projects in both China and the USSR, so acutely are new sources of supply needed.

1

From Population in relation to the development of energy from coal. Paper to the World Population Conference, Rome, August–September, 1954[1]

Future energy requirements will to some extent depend on the size of population; but the future size of population may also to some extent depend on the availability of energy. We are faced, therefore, with a 'circular problem', and in order to break the circle we must begin by making certain assumptions which can then be reconsidered in the light of the consequences to which they lead.

It will be convenient to start the investigation by taking the world as a unit and by considering, in the broadest fashion, the energy system as a whole. Any results obtained from the global approach can then be applied to particular regions and can be used for a study of that sector of the energy system which relates specifically to coal.

We begin, therefore, with an assumption regarding future world population. Present world population is taken to be about 2,500m. and world population in 1980 is assumed to be 3,000m.[2] This implies an increase during the next twenty-six years by much the same number as the increase since 1929. The assumed rate of increase, therefore, is considerably lower than that of the last twenty-five years.

It must be considered highly unlikely that a population increase of this magnitude could happen without an increase in the *per capita* use of energy. The population increase would presumably be associated with increased urbanisation and a substantial intensification of agriculture, both of which would make heavy demands on fuel supplies. A certain increase in the *per capita* use of energy would probably be needed simply to prevent living standards from falling. A further increase would be required to improve living standards. It is extremely difficult to quantify these factors. We are going to assume an average rate of growth in *per capita* energy *input* of 2 per cent cumulative — which means a 67 per cent expansion by 1980.

The assumed increase in population, combined with the assumed increase in the *per capita* rate of fuel requirements, yields a total increase by 100 per cent. In other words, world fuel requirements in 1980 would be twice as large as those of 1954.

The present pattern of world fuel consumption (measured on the *input* side) is roughly as shown in Table 2.1.1.

It is seen that roughly four-fifths of the fuel currently used by

TABLE 2.1.1

		%
'Capital Fuel'	Coal and Lignite	52
	Crude Oil and Gas	28
'Income Fuel'	Water power	1
	Wood	4
	Farm waste and dung	15
Total		100

man comes from 'capital' and only about one-fifth comes from current income. If total fuel requirements should double, it would (for many reasons) be unrealistic to assume an equivalent expansion in the 'current income' items of fuel supply. Nearly the whole of the additional load would have to be taken by the 'capital' items, i.e., coal, oil and gas, with possibly a contribution from atomic energy. On the basis of all the foregoing considerations (and a number of minor points which need not be presented here) we shall assume that the world requirement of coal and oil in 1980 will be at a rate that is $2\frac{1}{4}$ times the present rate.

The unit conventionally employed for dealing with fuel matters on a world scale is 'Q', defined as 10^{18} British Thermal Units.[3] Total world fuel consumption at present runs at the rate of roughly $0 \cdot 1$ Q per annum. On the assumptions stated, total world fuel consumption during the 26-year period up to 1980 would be roughly $4 \cdot 0$ Q, of which the 'capital fuels' − coal, oil and natural gas − would have to contribute about $3 \cdot 5$ Q. Experience during the last twenty-five years shows that world coal production, subject to increasing real costs, tends to take the place of 'residual supplier' in the world fuel economy. While coal output increased by only a few per cent, the output of oil/gas very nearly trebled. Practically the whole of the additional requirements of the world fuel economy were supplied by the 'liquid fuels', i.e., oil and natural gas. This preference for liquid fuels is continuing. Its continuation up to 1980 would mean that oil/gas consumption during the 26-year period would amount to $2 \cdot 0$ Q, while coal consumption amounted to $1 \cdot 5$ Q.

The same calculation carried forward to the year 2000 would indicate a total world fuel consumption (in forty-six years) of 9·7 Q, split up as shown in Table 2.1.2.

TABLE 2.1.2

Coal	2·5
Oil/Gas	6·5
Others	0·7
Total	9·7 Q

It will help to appreciate the magnitude of these figures and the revolutionary change of pattern, if we compare them with actual (estimated) fuel consumption during the last 100 years, split into two 50-year periods (see Table 2.1.3).

TABLE 2.1.3

	1855–1904	1905–1954	1955–2000
Coal	0·5	2·1	2·5
Oil/Gas	negligible	0·6	6·5
Others	0·6	0·6	0·7
Total	1·1 Q	3·3 Q	9·7 Q

We must now consider resources. Since coal, in actual historical experience, holds the position — more or less — of the 'residual supplier', we must necessarily begin with oil and natural gas. World consumption of oil/gas to date has been about 0·6 Q; proved reserves are stated to amount to about the same. Various authorities have given estimates of undiscovered oil reserves that might be recoverable at reasonable cost (thus excluding the bulk of the undersea oil reserves). These estimates agree on the order of magnitude — something like 5 Q. Total oil reserves, proved and unproved, recoverable at reasonable cost, therefore appear to be less than 6 Q (which, let it be remembered, would be roughly 10 times the total world oil consumption to date!). It would follow that world oil resources would be exhausted — except for very high cost residues — before the end of this century, if the assumptions of this paper were to become reality. In any case, there is every reason to believe that long before the end of this century a substantial proportion of the fuel load carried by natural oil/gas would have to be carried by synthesised liquid fuels produced from coal. Coal would then cease to be a mere residual supplier and would have to assume an ever growing importance.

World consumption of coal (including lignites) to date has been about 2·6 Q. World coal reserves, recoverable at costs compatible with the present economic structure of industrialised countries, may be taken as between 30 and 40 Q, of which roughly half would be located in the Soviet Union and China and the other half mainly in North America and Europe. Coal is not 'mobile' in the same way as oil: it is expensive to transport over long distances. Only a very small proportion of the world's coal production (1–2 per cent) enters world trade. It is therefore not realistic to use the global approach with regard to coal. But some very general conclusions can none the less be drawn.

On the assumptions here made, liquid fuel synthesised from coal would be required before the end of the century. If the oil/gas requirement is 6·5 Q, while total estimated oil/gas reserves amount to less than 6 Q, it might be reasonable to assume that only about 3 Q would be met from reserves, while 3·5 Q would have to be obtained through synthesis. As synthesis can hardly be expected to be achieved at more than 50 per cent thermal efficiency, the supply of 3·5 Q of liquid fuels synthesised from coal would mean an additional coal requirement of 7 Q. In other words, the total coal requirement during the 46-year period to the end of the century would be 9·5 Q, or about one quarter of the world's estimated coal reserves (recoverable at a reasonable cost).

Any further growth of world populations, combined with industrialisation on the pattern of Western technology, methods, and preferences, would then lead to the total exhaustion of 'economic' coal reserves throughout the world within another forty or fifty years.

We have already been at pains to emphasise that coal, unlike oil, cannot be treated as 'mobile'; the notion of world coal reserves in their relation to world fuel requirements, therefore, has no more than illustrative significance. If developments in population and *per capita* fuel consumption, such as given in our assumptions, were confined to the world outside the Soviet Union and China, and the fuel required for such developments were drawn without any contribution from those two countries, the situation would become critical in a much shorter time.

The 'global' analysis of fuel resources and requirements up to 1980 does not suggest that further substantial increases in world population during the next twenty-six years, combined with a 2 per cent annual expansion in *per capita* fuel requirements (measured on the *input* side), would necessarily 'run up' against an insurmountable barrier arising from the size of energy supplies. This conclusion is in no way affected by any assumptions anyone would like to make about the future of atomic energy; for atomic energy would in any case not become a significant factor in the world energy system by 1980.[4] Continuing expansion, beyond 1980, on the other hand, would undoubtedly encounter severe obstacles. If atomic energy came to relieve the fossil fuels, the 'obstacles' would be mainly financial (although the possibility of finding the raw material resources needed for the postulated atomic energy programme might also be open to doubt); if atomic energy did arrive 'in time', the 'obstacles' would be technical in an altogether final sense.

It is now desirable to leave the 'global' approach and turn attention to particular regions and populations . . . A few remarks must suffice. If the 'mobile' fuel, i.e., oil, is to take the main load of rising world fuel requirements, the fuel supply of specific regions or populations will depend mainly on their balance of payments situation. This cannot be analysed here at all. We therefore confine ourselves to coal. The world's coal resources, as is well known, are extremely unevenly distributed. Table 2.1.4 shows a very rough estimate of 'economical' reserves, contrasted with current rates of extraction. It is seen that the group called 'All others', which comprises about one-half of the world's population, disposed of only 4 per cent of the estimated world coal reserves and exploits them at a relatively fast rate. Even within this group the distribution of resources is extremely uneven; there are only four important coal producers, namely, South Africa, Australia, India and Japan; of these,

TABLE 2.1.4

	% of world reserves	% of world output
Europe (excl. USSR)	20	39
North America	26	36
Russia and China	50	16
All others	4	9
World	100	100

the former two account for about 2·9 per cent of world output and the latter two for about 5·2 per cent. Less than 1 per cent is contributed by 'the rest'.

These figures alone make it clear that coal can never in these countries become the energy basis for an economy developed on the pattern of Western industrialism. Even in Japan and India any large-scale industrialisation based on indigenous coal could only be short-lived. This does not mean, however, that any definite limit in the current rate of extraction has been reached today. Indian coal output (including Pakistan) has increased by 50 per cent during the last twenty-five years and could probably maintain a similar rate of expansion for another twenty-five years. Japan increased its coal output by 60 per cent in fifteen years (1929–44) and could presumably do something similar again. If Japan obtained access to China's coal her situation would, of course, be changed fundamentally.

It is inconceivable, within the existing framework of price relationships, that the inequality in the distribution of world coal reserves and output could ever be substantially mitigated by the large-scale movement of coal through international trade. As long as oil is freely available, coal will enter long-distance international trade only in exceptional cases; if oil should cease to be freely available, entirely new price relationships would establish themselves and coal, as the only concentrated fuel available without major capital investment, could then, at least in theory, again become a major factor in international trade. But we can leave that interesting possibility to the future.[5]

It is necessary to make a few concluding remarks. Owing to limitation of space, it has not been possible to mention every detailed aspect of the problem under discussion. Thus we have not specifically mentioned shale oil or tar sands as sources of fuel, nor have we dealt specifically with hydro-electricity, wind power, the harnessing of the tides, geo-thermic stations, vegetable sources, heat pump, photosynthesis, and so forth. The justification for these omissions lies in their quantitive unimportance which becomes obvious the moment one considers the matter from the economic, instead of the technical, point of view. It is true that the earth receives more solar energy than could ever be used by man. The economic difficulty is in collecting it at a cost which would even remotely fit into the present economic pattern. Nature has done this collecting job for man in the form of coal, oil and natural gas, and without any pretence to precision it can be said that industrial man at present burns within a year what it has taken Nature a million years to accumulate. It is not surprising therefore that any attempt to maintain a similar rate of consumption from energy collected by man himself, as he goes along, should turn out to be somewhat expensive. We are of course aware of the possibility that 'something might turn up'. This expectation, supported by entirely premature hopes fastened on atomic energy, is likely to stand in the way of a realistic appreciation of the energy problem to be faced by industrialism on the 'Western' pattern.

2

From Prospect for coal. Address to a conference of university appointments officers, April 1961[1]

With regard to future oil supplies, as with regard to atomic energy, many people manage to assume a position of limitless optimism, quite impervious to reason.

I prefer to base myself on information coming from the oil people themselves. They are not saying that oil will shortly give out; on the contrary, they are saying that very much more oil is still to be found than has been found to date and that the world's oil reserves, recoverable at a reasonable cost, may well amount to something of the order of 200,000m. tons, that is about 200 times the current annual take. They leave it at that, and I, of course, accept what they say. Let us try, however, to see what these figures mean. We know that so-called 'proved' oil reserves stand at present at about 40,000m. tons, and we certainly do not fall into the elementary error of thinking that that is all the oil there is likely to be. No, we are quite happy to believe that the almost unimaginably large amount of a further 160,000m. tons of oil will be discovered during the next few decades. Why almost unimaginable? Because, for instance, the great recent discovery of large oil deposits in the Sahara (which has induced many people to believe that the future prospects of oil have been fundamentally changed thereby) would hardly affect this figure one way or another. Present opinion of the experts appears to be that the Saharan oil fields may ultimately yield as much as 1,000m. tons. This is an impressive figure when held, let us say, against the present annual oil requirements of France; but it is quite insignificant as a contribution to the 160,000m. tons which we assume will be discovered in the foreseeable future. That is why I said 'almost unimaginable', because 160 such discoveries as that of Saharan oil are indeed difficult to imagine. All the same, let us assume that they can be made and will be made.

It looks, therefore, as if proved oil reserves should be enough for forty years and total oil reserves for 200 years — at the current rate of consumption. Unfortunately, however, the rate of consumption is not stable but has a long history of growth at a rate of 6 or 7 per cent a year. Indeed, if this growth stopped from now on, there could be no question of oil displacing coal, and everybody appears to be quite confident that the growth of oil — we are speaking on a world scale — will continue at the established rate. Industrialisation is spreading right across the world and is being carried forward mainly by the power of oil. Does anybody assume that this process would suddenly cease? If not, it might be worth our while to consider, purely arithmetically, how long it could continue.

What I propose to make now is not a prediction but simply an exploratory calculation or, as the engineers might call it, a feasibility study. A growth rate of 7 per cent means doubling in ten years. In 1970, therefore, world oil consumption might be at the rate of 2,000m. tons per annum. The amount taken during the decade would be roughly 15,000m. tons. To maintain proved reserves at 40,000m. tons new provings during the decade would have to amount to about 15,000m. tons. Proved reserves, which are at present forty times annual take, would then be only twenty times, the annual take having doubled. There would be nothing inherently absurd or impossible in such a development. Ten years, however, are a very short time when we are dealing with problems of fuel supply. So let us look at the following ten years leading up to about 1980. If oil consumption continued to grow at roughly 7 per cent per annum, it would rise to about 4,000m. tons a year in 1980.[2] The total take during this second decade would be roughly 30,000m. tons. If the 'life' of proved reserves were to be maintained at twenty years — and few people would care to engage in big investments without being able to look to at least twenty years for writing them off — it would not suffice merely to replace the take of 30,000m. tons; it would be necessary to end up with proved reserves at 80,000m. tons (20 times 4,000). New discoveries during that second decade would therefore have to amount to not less than 70,000m. tons. Such a figure, I suggest, already looks pretty fantastic. What is more, by that time we would have used up about 45,000m. tons out of our original 200,000m. tons total. The remaining 155,000m., discovered and not-yet-discovered, would allow a continuation of the 1980 rate of consumption for less than forty years. No further arithmetical demonstration is needed to make us realise that a continuation of rapid *growth* beyond 1980 would then be virtually impossible.

This, then, is the result of our 'feasibility study': if there is any truth at all in the estimates of total oil reserves which have been published by the leading oil geologists, there can be no doubt that the oil industry will be able to sustain its established rate of growth for another ten years; there is considerable doubt whether it will be able to do so for twenty years; and there is almost a certainty that it will not be able to continue rapid growth beyond 1980. In that year, or rather around that time, world oil consumption would be greater than ever before and proved oil reserves, in absolute amount, would also be the highest ever.[3] There is no suggestion that the world would have reached the end of its oil resources; but it would have reached the end of oil growth.

3

From Future energy prospects. Paper to National Coal Board management conference, Scarborough, April 1962

If fuel is short, everything is short. Although one fuel can substitute for another, nothing can substitute for fuel. The individual consumer may indeed decide to spend less of his money on fuel, so as to be able to spend more on other things; but as the other things themselves are made with fuel, there is no likelihood that the total use of fuel is reduced thereby.

Too many people talk about fuel as if it were just another commodity, to be judged and dealt with in exactly the same way as any other goods or services offered to the consumer. That is to say, they refuse to consider anything but the momentary situation and the immediate prospects, whereas what matters most when dealing with fuel are the longer-term prospects, the situation as it is likely to evolve, not over the next few years, but over the next few decades.

The world's fuel consumption in 1924 amounted to about 1,500m. tons of coal equivalent. At present, it runs at a level more than three times as high — over 4,500m. tons a year, annual consumption per head averaging at about $1\frac{1}{2}$ tons. What are we to expect for the year 2000, thirty-eight years hence? There is a world-wide movement of industrialisation and urbanisation, supported by massive aid from the rich countries to the poor. Are we to assume that the per capita rate of fuel consumption will *not* be affected by these developments? Or that it will increase as it has increased in the past, or that it will increase somewhat faster? What I have to say is not highly sensitive to the assumption we may choose to adopt. No one could believe that the poor will be content to remain as they are, or indeed that the rich will be content to leave them where they are. The combined effect of the increase in world population and even a modest increase in the per capita rate is certain to be prodigious. If the per capita rate should increase only slightly faster as it has done in the past thirty-eight years, total fuel use in the year 2000 will be about four times as high as it is now, and ten times as high as it was in 1924 — something of the order of 18,000m. tons of coal equivalent a year.

What do the United States, the greatest oil producer and consumer in the world, think about the prospects for their coal? The Bureau of Mines estimate that US coal consumption, currently around 400m. tons per annum, will rise to 670m. tons by 1975. There is some doubt whether output could be stepped up to that level in thirteen years.[1]

It is not my task to speculate whether world fuel requirements at the rate of 18,000m. tons a year could ever be met. The world's proven oil reserves, at about 40,000m. tons, pale into insignificance as against an annual requirement of that order. Even if we accept the estimates of the world's leading oil geologists, which suggest that total recoverable oil reserves yet to be discovered might be of the order of 250,000m. tons, it is obvious that mankind will need, in the foreseeable future, immense amounts of fuel in addition to all the oil it can hope to extract.

The use of oil, like that of electricity, tends to double every ten years. A sober appraisal of the oil reserves estimates inevitably leads to the conclusion that the annual rate of oil consumption will probably have reached its peak when it has risen fourfold from the present level, which may have come to pass by the early 1980s.[2] In about twenty years' time [1982] the hunger of the underdeveloped countries, whose populations also show the greatest rates of increase, will by no means be satisfied; on the contrary, the demand for speedy economic expansion will probably be greater than ever before. But how is economic expansion to be carried forward without increasing supplies of oil? In other words, the critical moment for the world's oil economy arrives, not when oil reserves have been finally exhausted, which may be many centuries off, but when the annual rate of oil supply has reached its maximum and cannot be raised any further. At that time, obviously, proven oil reserves will be greater than ever before in history — yet not great enough to sustain further expansion of the annual output.

4

From Coal in the fuel economy. Paper to the Fourth
International Mining Congress, London, July 15, 1965[1]

Since the end of the war, the world economy has been in a state
of breathless expansion. Nothing like it has ever been seen
before. The best picture of this expansion, I think, is obtained
by looking at world steel production. Steel is not an article of
direct consumption, but it enters almost every kind of economic
activity and therefore gives a good indication of total activity.
As an important element of economic life, it is barely 100 years
old. The world steel industry reached an output of 100m. tons a
year after about seventy years of strenuous expansion, in 1935.
Sixteen years later, it reached 200m. tons; eight years later,
300m. tons; five years later, over 400m. tons. In the last fifteen
years, the world has produced and used more steel than in its
entire previous history — over $4\frac{1}{2}$ milliard tons.[2]

It is against this background of world-wide expansion that we
must consider the world's fuel resources and the role of coal
among them. Coal has been the main basis of economic
development in all major industrial countries. It used to be
indispensable. Is it still indispensable?

The coal industries of many countries are making immense
and highly successful efforts to improve safety, raise productivity, and save manpower. They stand at the beginning of a
technical revolution which holds the highest promise. But what
if other fuels are going to expand so rapidly that the world is not
going to need coal — or at least so much coal? Then the mining
industries are wasting their time. It is pointless to keep alive
something that has been superseded.

This, then, is our question: Are we moving into an age
without need for coal, a coal-less age? Can we now say goodbye
to coalmining, that costly and dangerous operation, because
natural gas and oil will gush into our hands in such volumes and
at so little effort that it would be folly not to use them for all the
purposes currently served by coal; or because nuclear energy in
virtually unlimited quantities is just around the corner?

As such questions are being asked, we must try to answer
them.

The expansion of world economic activity, which I have illustrated by referring to steel, has up to now increased the demands for every kind of fuel, including coal, the biggest supplier of all. The share of coal has indeed declined, but it is quite wrong to think that the volume of coal has fallen; it is today bigger than ever before. Coal output is growing at a lower rate than oil output, but in absolute quantity it has recently been growing just as fast, and the lower rate of growth is merely a reflection of the coal industry's still being bigger than the oil industry. Thus, from 1962 to 1963, coal output expanded by about 130m. tons, and oil output by about 130m. tons of coal equivalent.

To supply the world's fuel needs is so large an operation that it is hard for the human mind to grasp. To illustrate : just over one third of the world's fuel supplies consists of oil, and about one half of this oil is carried across the seas. Yet the sea-going oil accounts for more than one half of all sea-going transport. In other words, to transport about one sixth of the world's fuel across the ocean is a bigger job than the transport of all other goods combined. Or another illustration: to grow at the rate of 4–5 per cent a year, the oil industry claims to require an amount of capital which is equal to all that annually made available on the capital markets of Europe and America. As the industry considers it to be unlikely that it could raise in the capital markets more than about one tenth of the required amount, it will have to cover about 90 per cent of its capital requirements by self-financing.

At present, total world fuel requirements exceed 5,000m. tons of coal equivalent, of which coal (including lignite) provides about 2,500m. tons. Because of world economic expansion, total fuel needs are rising fast. They rose by about 70 per cent during the last ten years and are generally expected to double in the next fifteen years or so. This would bring total requirements into the region of about 10,000m. tons in the early 1980s.[3]

An important part of this growth, of course, is attributable to the growth of world population. But living standards are growing as well, and expectations are growing even faster. If we convert the world's fuel use into 'manpower equivalents' or 'mechanical slaves', we might say that, on the average, every

44

living person today has six 'mechanical slaves' to help him. But this average hides great disparities. The United States of America, for instance, can boast 33 'mechanical slaves' per head of the population, while the so-called developing countries have barely one. Even in the United States no one is speaking of saturation, so that it is fair to conclude that the potential growth in the world's fuel use is practically unlimited. A continuation of the rate of growth experienced during the last twenty years would carry world fuel needs to the staggering total of about 20,000m. tons of coal equivalent by the year 2000 — in thirty-five years' time.

The question of whether the world will be able to cover its fuel requirements without drawing heavily on coal must sound a strange one to anyone who has comprehended these magnitudes. If total supplies have to double in fifteen years or so, is it likely that this can be accomplished while neglecting coal, which at present accounts for about one half of total supplies? What is there to take the place of coal and at the same time to look after the requirements of growth?

Fossil fuels — coal, oil and natural gas — are won from nature's reserves, and the size of these reserves is obviously important in our consideration of the question of the future of coal. Estimates of 'ultimate reserves' are, of course, extremely uncertain. Even if the geologists had complete and perfect knowledge of the composition of the earth's crust, they would still have to introduce an economic criterion, a judgment as to which reserves, at what depth and of what quality and concentration, could be reasonably included in the estimate. All the same, certain estimates were given at a recent World Power Conference which are worth quoting (see Table 2.4.1).

These estimates may well be wildly wrong; but it is clear that even very substantial revisions of the figures quoted for oil, natural gas, and shale oils would not alter the main message — that if mankind continues to depend largely on fossil fuels in the long run, it will depend on coal rather than oil or natural gas.

However that may be, it can well be argued that the long run is of less interest than the short run, and that even on these figures there should be plenty of oil and natural gas 'for the foreseeable future'. 'Proved Reserves' in relation to current output are more relevant than 'Ultimate Reserves'.

The world's proved oil reserves have risen tremendously during the last quarter of a century — from about 4·6 milliard tons to nearly 46 milliard tons, a factor of ten. Meanwhile, oil

TABLE 2.4.1   *World Reserves of Fossil Fuels*

	Million million tons of coal equivalent	%
Hard Coal	4·1	66
Lignite	1·0	16
Peat	0·8	13
Sub-total	5·9	95
Oil	0·2	3
Natural Gas	0·05	1
Shale Oils	0·05	1
Sub-total	0·3	5
TOTAL	6·2	100

production rose from 268m. tons in 1938 to 1,400m. tons in 1964, a factor of just over five. This would seem to be a highly reassuring picture: more and more is being taken and yet proved reserves grow at a faster rate than the take. The picture is less reassuring when we look at the more recent past. In 1958, the relationship between proved reserves and the year's output — which we may call the 'life' — was 41 years. By 1964, the 'life' had fallen to 32·6 years. In other words, during the last six years the 'life' of proved oil reserves has declined by nearly 8½ years; the growth of proved reserves has not kept pace with the growth in output.[4]

5

From The place of coal in the future. Address to Combustion Engineering Association Conferences in Manchester, January 25, and London, February 17, 1966[1]

The Americas have resources for self-sufficiency of a very high level of fuel consumption. The Soviet Bloc can be self-sufficient at an increasing level of consumption, although they may get into temporary difficulties if fuel production does not quite keep in step. The under-developed world has a very low level of fuel consumption and is also poorly endowed with fuel resources. There is the one great surplus area — the Middle East, and if one looks in detail into the political constitution, the history, and geographical position of this area, it is quite clear that it is the most uncertain and unreliable area of the world, yet Western Europe, including the United Kingdom, every year increases its dependence on oil supplies from the Middle East, and appears to be prepared, for a temporary benefit of low fuel prices, to destroy its only indigenous fuel base, which is coal.

However, even if we leave this factor of insecurity on one side, as we should not, and look at the quantities involved, fully realising that primary fuel is not produced by man but is found where the Lord has put it: what do we find? Oil production started just 106 or 107 years ago. Since then the total consumption of oil has been 142 milliard barrels, an enormous figure and beyond our powers of imagination. However, with world development plans and expansionist trends, which all rely mainly on oil, it is estimated that in the next twelve years the total consumption will be 150 milliard barrels, equal to somewhat more than the consumption during the entire history of the oil industry. This means that to maintain a proper reserve ratio two more regions with an endowment similar to that in the Middle East will need to be discovered in the next twelve years. In the last fifty years only one such region, the Middle East, has in fact been found and while I am not saying that it is unthinkable that two more will be found in the next twelve years, I am presenting this as a problem.

6

From 'Coal and its competitors', *Colliery Guardian Annual Review,* 1968[1]

It is not very clear how natural gas could act as an effective barrier against a rise in oil prices. On a world-wide scale, it does not really compete with the oil that flows into international trade. And even as regards the United Kingdom, North Sea gas is only a limited energy source, expected to contribute not more than 15 per cent of Britain's energy in 1975, if all goes well.[2] At the present time, North Sea gas reserves are insufficient to do even that; but more gas may well be found. It would be quite unrealistic to imagine that enough gas will be found to replace the bulk of UK oil consumption, in case there should be a sharp rise in oil prices. In short, oil prices cannot be 'held in check' by natural gas; if oil prices rose sharply, the producers of natural gas would simply make higher profits, most of which would accrue to foreign companies and thus burden the balance of payments.

7

From 'Western Europe's energy crisis — a problem of life-styles', *Ambio*, Royal Swedish Academy of Sciences, 1973[1]

It is impossible to exaggerate the dependence of the modern world on fossil fuels. As these fuels are non-renewable and constitute a once-for-all endowment of the earth, their availability in terms of quantity, and therefore also of time, is limited, and it must give rise to increasingly serious concern that the modern economy seems to be inexorably geared to a continuous, exponential growth in its requirement for them. In 1971 — the latest year for which global statistics have been assembled — total world energy consumption amounted to 7,260m. tons of coal equivalent, nearly 300m. tons more than in 1970, and of this total oil and natural gas accounted for 64 per cent, solid fuel for 34 per cent, and primary electricity (i.e. hydro-electricity and nuclear energy) for 2 per cent. Thus fossil fuels, which strictly speaking are non-reproducible 'capital', supplied 98 per cent of all requirements, while reproducible 'income fuels' supplied only 2 per cent. It can be argued that this understates the contribution of the latter; by the application of different conversion factors, 'income fuels' could be shown to contribute 6 per cent of the total. But this does not alter the basic situation of virtually total dependence on 'capital'.

The geographical distribution of fossil fuel reserves as well as that of fuel demand, exacerbates the situation. From the point of view of demand versus resources, the world may be divided into four groups of countries:

Group   I with high consumption rates and large indigen-
            ous resources
Group II with high consumption rates and small resources
Group III with low consumption rates and large resources
Group IV with low consumption rates and small resources

The outstanding representatives of Group II are Western Europe and Japan, while the United States of America is rapidly moving into a similar position. The outstanding representatives of Group III are the Middle East and North Africa. The relationship between these two groups is of crucial importance − at least in the short run of the next ten to thirty years. The countries of Group III are increasingly becoming aware of their immense bargaining power − which they have solidified by setting up OPEC, the Organisation of the Petroleum Exporting Countries, and they are also becoming aware of the enormous difficulties encountered in developing an alternative livelihood for their populations, to be available when their oil reserves run out. Having discovered that the demand for their oil is virtually inelastic even at steeply rising prices, they can obtain more money by releasing less oil, and many of them already find themselves with a greater foreign exchange income than they can spend on imports. What are they to do with their cash surpluses? They are gradually coming to the conclusion that the best long-term investment for them may be to leave the oil in the ground and release it only at the rate determined by their foreign exchange needs. These developments, so far gradual but none the less implacable, are

49

placing the countries of Group II in a position of immediate danger.

Even if there is no question of a total and permanent stop in the flow of oil supplies from the OPEC countries, a temporary interruption due to political factors, or a significant permanent reduction could play havoc with the economies of Group II countries and cause a degree of paralysis and disorder from which it might be extremely difficult to recover.

The question needs to be asked: What accounts for the extraordinarily high level of fuel and energy consumption in the so-called advanced countries, while traditional societies seem to be able to get along with very small fuel inputs indeed? The answer normally given simply refers to the higher level of income of the former. But this answer is too superficial. Take, for instance, the production and consumption of foodstuffs. In 'advanced' societies the production process requires fuel and energy inputs which are a high multiple of those required by traditional societies, although food consumption levels are not very much higher in the former than in (some of) the latter. A study of these situations provides some valuable pointers. Similar pointers can be obtained from historical analysis. During the last twenty-five years, for instance, the fuel requirements of agriculture in the advanced countries, including the fuel requirements of agricultural inputs as well as those of food processing, have increased by a far higher factor than the increase in agricultural output. Why is it, then, that agriculture – per unit of output – has become so enormously more 'fuel dependent'?

The answer can be found only in the emergence of:

new patterns of production
new patterns of consumption
new patterns in the geographical distribution of populations

American studies have shown the enormous dependence of American agriculture on fossil fuels.

Harvested crops capture solar energy and store it as food or

50

some other useful product. Yet the energy captured is small compared to the energy we burn to capture it. Agriculture, as a result, has become a major consumer of our stores of energy, using more petroleum than any other single industry. If the world is facing a future with rising energy prices, the highly mechanised technology currently used in US agriculture may be inappropriate.*

Professor Barry Commoner has given some telling figures:

> In 1949, an average of about 11,000 tons of fertiliser nitrogen were used per . . . unit of crop production, while in 1968 about 57,000 tons of nitrogen were used for the same crop yield. This means that the efficiency with which nitrogen contributes to the growth of the crop declined fivefold.†

During the same period, the US population increased by 34 per cent; total US agriculture production rose by 45 per cent; the annual use of fertiliser nitrogen increased by 648 per cent; and the harvested area declined by 16 per cent.‡ During a similar period, between 1946 and 1971, the proportion of the American working population engaged in agriculture fell from 14 per cent to 4·4 per cent. It is in these changes in the pattern of production, not in any inefficiency of fuel utilisation in mechanical equipment or chemical factories, that an explanation of agriculture's enormous dependence on fossil fuels must be sought.

The patterns of consumption have been changing in a similar way. In the advanced countries, very little food reaches the consumer in its natural state. Virtually all foodstuffs are elaborately processed, packaged, and transported over long distances. With the growth of specialised mass production, consumption has become widely separated from production – and all this requires prodigious amounts of cheap mechanical and process energy.

What are the possibilities of developing alternative sources of fuel supply to fill the gap if oil should become scarce? When this question is raised, people are inclined to point to the allegedly unlimited possibilities of nuclear energy. As long, however, as no method exists for the safe disposal of radio-activity, any large-scale development of fission energy would be nothing short of suicidal. Such a large-scale development, it must also be emphasised, would be possible only on the basis of breeder-reactors. However, 'the expected switch to fast breeder reactors will aggravate the situation even further, for they produce large quantities of radio-active substances with very long half lives'.§ It has also been observed that 'one of the most disturbing features of nuclear generators in current use is that exhausted reactor cores cannot be dismantled, but must be sealed and buried . . . People are seriously worried about risks of exposure to radiation, either through accidents of unimaginable dimensions or through the cumulative effect of small doses, directly experienced or indirectly transmitted.'** And 'breeder reactors are inherently more difficult to control because the process takes place much faster'†† than in any of the reactors now in use. In view of these appalling dangers, there is now a good deal of talk about the limitless possibilities of fusion energy. 'Generation by nuclear fusion produces no radio-active wastes . . . The difficulty is that the process only works continuously at enormously high temperatures (up to 200m.°C).'‡‡ What the massive production of sun temperatures on earth would do to the living environment is completely unknown; but, in any case, the practical feasibility of such a process applied on a quantitatively relevant scale has in no way been established, so that no realistic policies can be based on the expectation of its successful implementation within the foreseeable future.

The discussion therefore turns to the possibility of the large-scale utilisation of solar energy and its derivatives, and also of tidal power and geothermic heat. These sources of 'income energy' are, of course, very large, inexhaustible (with some reservations regarding geothermic heat), and extremely widely spread over the globe. The difficulty is that, being widespread, they are inherently diffuse and cannot easily be concentrated or centralised into large and continuous supplies, such as the modern world is used to with fossil fuels. The question of solar energy therefore immediately raises the question of 'life-style' . . . Highly diffuse energy would fit only a highly decentralised mode of living. We are therefore brought back to the proposition that the modern world's dependence on fossil fuels could be significantly reduced only by means of the development of new 'styles of technology'.

8

From *Facing the change in the energy situation*. Address to a conference organised by the Royal Institute of British Architects, University of Durham, July 18, 1974[1]

Modern agricultural technology, as practised in the United States, in Western Europe, in the areas affected by the 'green revolution', and in many other parts of the world, is essentially oil-based. Its tremendous success in raising productivity-per-man was achieved by the introduction of an intensely oil-based technology — intensive mechanisation and, even more importantly, intensive chemicalisation. In terms of physics and chemistry, modern man eats a variety of foodstuffs; but in terms of economics, he eats oil.

As we have already seen, the modern pattern of urbanisation would have been impossible without an unparalleled rise of productivity-per-man in agriculture, which, in turn, would probably not have happened without oil-based mechanisation and chemicalisation. This modern system of agriculture is obviously extremely vulnerable to adverse changes on the oil front. We can see this already today as we witness the virtual collapse of the 'Green Revolution' in a number of developing countries. It has been calculated how much oil would be required if the whole of mankind — some 4,000m. people, forgetting any further increases in world population — were to be fed by means of agricultural technology. The answer is that on such assumptions all proved oil reserves, as currently known, would be exhausted *by agriculture alone* within less than thirty years.

3
Exploratory Calculations into Energy Consumption

Introduction

About the only thing that disturbed Schumacher's sunny equanimity was the activity of economic forecasters who confidently predicted, sometimes with great precision of detail, what the future demand for every form of energy would be. Yet in a business with such long lead times for investment, it was necessary to attempt some estimates as a broad basis for policy.

His method of making exploratory calculations proved much more effective and the estimates he made in 1963 of likely world demand for energy in the year 2000 would still be approved by most authorities now that we are half way between the two years. His scepticism about the forecast rate of growth in nuclear power has certainly been justified (the United States Department of Energy has now halved its expectations of its contribution in 1990), as have his doubts about the economic arguments used to support the reactor building programme, especially if the research and development costs are taken into account. Sadly, his misgivings about the effect of the coal industry's rundown on relations with the mineworkers and their Union also proved well founded by two national strikes in the early 1970s.

I have been able to find in all the material he produced on energy only one misjudgment: in 1952 he argued that there were no great new areas of coal to be exploited in Britain. In fact, when the Coal Board's exploration programme was greatly increased as a response to the 1973 Middle East crisis, two major coalfields in virgin areas (Selby in Yorkshire and North East Leicestershire) were proved and almost certainly there are more to come.

1

Review of *Étude sur les perspectives énergétiques à long terme de la Communauté européenne*[1] in the *Economic Journal*, Cambridge University Press, March 1964

This study provides answers to almost any economic question anyone might wish to ask about fuel and energy in the Common Market countries up to 1975. What will be the growth of the national product between 1960 and 1975? – 99 per cent. And the growth of industrial production? – 136 per cent. Of steel output? – 78 per cent. And the concomitant growth of total fuel consumption? – 84 per cent. Of fuel consumption per head? – 65 per cent. Of electricity? – 176 per cent. What will be the increase in fuel efficiency at power stations? – 23 per cent. Etc., etc. But this is not all. We may go along with the extrapolation of trends relating to physical quantities, even though experience warns us not to take them very seriously; but can we still go along when we are given figures about future wages, prices, productivities and costs? What will be the average output per man underground in each of the seven main coalfields of the Community in 1975? – South Belgium will be the lowest at 2,390 kg., while Lorraine will be the highest at 4,220 kg.

No doubt there are compelling reasons for estimating that, for instance, the average rate of increase in French miners' wages will be 4·0 per cent in the first quinquennium, 3·8 per cent in the second and 3·9 per cent in the third. Unfortunately, these reasons have not been disclosed.

It may seem astonishing enough that anyone should be able to predict the development of miners' wages and productivity in his own country fifteen years ahead: it is even more astonishing to find him predicting prices and transatlantic freight rates of American coal. A certain quality of U.S. coal, we are told, will cost 'about $14.50 per ton' free North Sea port in 1970, and 'a little more' in 1975. 'About $14.50', the report says, should be taken as meaning 'anything between $13.75 and $15.25', a margin of uncertainty of $1.50 or ± 5 per cent. Similarly, the price of fuel oil in 1975 will be something of the order of $17–19 per ton, while estimates of various kinds are given for natural gas and nuclear energy. Being in the possession of these (and many other) 'facts', the authors find it an easy matter to calculate how much of the Community's coal production will be competitive in 1970, and the answer is 'about 125m. tons, i.e. a little over half the present production'.

The question is: What is the value of such a study? What is the meaning of the forecasts presented in such astonishing detail? Are they likely to be 'true', at least sufficiently 'true', to serve, as they are intended, as a basis for policy making?

Let us look closely to see how these forecasts are manufactured. Take transatlantic freight rates, which are, of course, a decisive determinant of the competitive position of U.S. coal in Europe. Within just over three years — the beginning of 1956 and the middle of 1959 — they dropped from $10 to $2.90 a ton. During the Suez crisis at the end of 1956 they had risen to $15 a ton. The authors admit that these figures make it appear impossible to arrive at a meaningful forecast for 1970 or 1975. But they offer a forecast all the same — $3.50 to $5.00 a ton — on the basis, they say, of a study of the operating costs of such ships as may be constructed during the next few years.

Similarly, the pithead price of U.S. coal — another decisive element. The authors recognise that U.S. coal consumption may rise to 750m. tons (from about 400m.) by 1975. Such a change would seem to make all forward cost estimates exceedingly hazardous. None the less, we are told that the pithead price of steam coal will rise from $4.40 in 1960 to $4.65 in 1965 and 1970.

I think the time has come when it is necessary to be quite blunt about this sort of thing: these figures are not worth the paper they are written on. They are a case of spurious verisimilitude which borders on mendacity. The whole study, from beginning to end, proceeds from tendentious assumptions to foregone conclusions. It does not add one iota to knowledge, at the same time doing the greatest disservice to the advancement of knowledge by offering arbitrary imaginings as if they were respectable results of a scientific inquiry.

It has rightly been said that the beginning of wisdom is to know the limits of one's knowledge. To the question: 'What are transatlantic coal freight rates likely to be in 1975?' there is only one honest scientific answer: 'I do not know'. And the same is true of most of the other questions asked and answered in this study.

Honesty is not restored by repeated references to the uncertainty of the estimates; nor by the introduction of 'ranges' instead of firm figures. The ranges themselves are guesswork and merely add to the pretence. It is not a matter of uncertainty but of arbitrariness. If I put down a figure as my estimate, however uncertain I may be I am asserting that in my view this figure is more defensible than any other. The simple truth is that when it comes to prices, wages and many other economic factors in 1975 all figures, within a very wide range, are equally defensible. Nobody knows, and nobody can possibly know, whether (e.g.) U.S. coal in 1975 will be competitive with indigenous coal in Europe or not. A study which comes to the conclusion, as this one does, that 'la position concurrentielle des charbonnages ne s'améliorera pas à long terme' (p. 196) and that 'les quantités de charbon communautaire compétitives en l'absence de toute aide ne seraient que légèrement supérieures à la moitié de la production actuelle' is scientifically and morally indefensible — not because it comes to an unfavourable conclusion for the Continental coal industries, but because it pretends to offer knowledge where there is none.

It is fashionable today to assume that any figures about the future are better than none. To produce figures about the unknown, the current method is to make a guess about something or other — called an 'assumption' — and to derive an estimate from it by subtle calculation. The estimate is then presented as the result of scientific reasoning, something far superior to mere guesswork. This is a pernicious practice which can only lead to the most colossal planning errors, because it offers a bogus answer where, in fact, an entrepreneurial judgment is required.

The study here under review employs a vast array of arbitrary assumptions, which are then, as it were, put into a calculating machine to produce a 'scientific' result. It would have been cheaper, and indeed more honest, simply to assume the result.

2

From Population in relation to the development of energy from coal. Paper to the World Population Conference, Rome, August–September, 1954[1]

It is necessary to stress again the purely 'exploratory' character of the calculations presented in this paper. To formulate certain more or less conventional assumptions and to trace their effect over a period of time is something very different from prophesying. There is here no intention whatever to predict that oil/gas supplies will *in fact* run out before the end of the century, or that coal supplies will *in fact* run out by the middle of the twenty-first century. Developments of this kind never proceed in such a clear-cut manner; they set up new forces of a generally unpredictable nature. But the 'exploratory' calculation helps to determine the magnitude or 'intensity' which any 'new forces' would have to possess and thus to judge to what extent the 'new forces' themselves would be compatible with existing patterns and tendencies in general and with our specific assumptions in particular.

It is . . . highly probable that natural oil/gas, in spite of its relatively short life expectancy, will be able for another twenty-five years or so to compete successfully with coal and *to determine, on the basis of its own production costs, the general level of fuel prices.* The 'elasticity of supply' of oil/gas is far greater than that of coal, as all experience over the last fifty years has demonstrated. In other words, the fact that coal will probably have to come into ever greater prominence after 1980 does not mean that the coal industry might not have a very difficult stand up to 1980. The optical illusion thus created could easily lead to a gross neglect and wasteful exploitation of the world's coal resources during the intervening period, coupled with the belief that coal resources were 'inexhaustible' inasmuch as current rates of extraction showed no marked tendency to rise.

3

From Nuclear energy and the world's fuel and power requirements. Address at a conference on nuclear energy organised by the Federation of British Industry, April 10, 1958

I want to make it quite clear that I am not giving you forecasts, not even 'statements of the most probable course of events'. I am going to engage in certain 'exploratory calculations', that is all.

It will be useful, I suggest, simply to explore, with the help of ordinary arithmetic. I do not know — and I suggest nobody knows — at what rate, or for how long, world fuel requirements will rise in the future. But we can work out what certain assumed rates of increase would mean. For instance, some people feel that the present American way of life is now, or should be, the birthright of anyone born on this planet, and that 'technical assistance' should bring it about as soon as possible that everybody will live the American way. Without discussing the desirability or otherwise of such a development, we can at once make an exploratory calculation: assume the U.S. way of life at present requires about eight tons of coal equivalent per person per year. So the fuel requirements of today's world population at an American standard would be about 20,000m. tons of coal equivalent a year, or roughly six times the present. If an appreciable part of this were to be taken in the form of oil (it has been pointed out*) 'all potentially recoverable petroleum in the earth would probably be consumed in a very few years'.

What is the value of this simple and all-too-obvious calculation? To show that the American way of life cannot be spread over the whole world, except possibly at a very much higher level of technology than the present American level. It can equally easily be demonstrated that the same applies to the much more modest British way of life, in fact to the Western industrial system as such. There is certainly room for a good deal of 'development of backward areas'; but this development, if it were to lead to European or American rates of fuel and power consumption, could not be sustained for long on the basis of present-day techniques.

Let us go further and make a rough exploratory calculation bringing in the factor of time. Could the whole world attain, let us say, half the present American standard by the year 2000? Assuming that the Americans themselves stabilise their own consumption at the present per capita level, and assuming the world population will rise to 5,000m. by the year 2000, total fuel consumption would have to rise to about 21,000m. tons of coal equivalent a year, or much the same as if the entire present world population were brought up to the American standard at once. To effect such a six-fold increase in just over forty years would require an average annual rate of growth of $4\frac{1}{2}$ per cent compound for all fuels taken together. Since the production of coal is relatively inelastic, the growth of oil production would have to be appreciably higher, say 8–9 per cent a year compound. A growth rate of that magnitude is not unknown in the oil industry and would, in fact, represent just a continuation of a trend established over the last twelve years. However, if continued for only a few decades, all the Earth's potential reserves of natural oil would be totally exhausted before the world had attained the desired standard, namely, one-half of the present American level.

Nor do we obtain more plausible results if we allow a longer period for the attainment of such standards, e.g. to the year 2050. By that time, it is estimated, the world population will have risen to about 7,000m. If the North American population is allowed no more than its present per capita consumption, while the rest of the world is allowed one-half of that standard, total requirements in 2050 would be about 30,000m. tons of coal equivalent. The rate of growth required to attain this level at that time would be just less than 2·5 per cent compound per annum, or much less than the actual rate of growth of commercial fuels over the last hundred years. Assuming that atomic energy will supply one-third of the total, while coal and water supply three times as much as at present, oil supplies would have to amount to about 10,000m. tons per year. The highest estimates ever given of the world's potential reserves of oil, gas, oil shales and tar sands fail to yield a figure sufficient to sustain the assumed development up to the year 2050.

These are simply exploratory calculations, testing certain assumptions by means of straightforward arithmetic. They are in no way forecasts; in fact, they may well be called anti-forecasts because they demonstrate that certain sets of assumptions do not add up to a plausible result.

4

From Long-term demands for fuel. Paper to a study group of the Royal Statistical Society, May 21, 1958[1]

Before the Second World War, studies dealing with the long-term demand for fuel were exceedingly rare. Fuel was something that everybody was inclined to take for granted; in the inter-war period it tended to be over-supplied rather than scarce; and there was hardly an occasion for attempting to 'extrapolate' demand curves, particularly since an extrapolation of 'no change' is in any case somewhat uninteresting. In the second half of the nineteenth century, there was Jevons's book, *The Coal Question,* written during a period of rapid growth in the production and consumption — both inland and abroad — of British coal; it was first published in 1865.

Jevons's book is nowadays often cited as an awful warning against long-term forecasting. The warning is no doubt justified, but the implied criticism of Jevons is quite wrong. The book, in its main thesis, does not contain any forecasts at all; it contains what I propose to call 'exploratory calculations'. Having observed a certain rate of growth over several decades, Jevons shows how great the consumption of coal would become if the established rate of growth were maintained. Far from prophesying that it will be maintained, he argues, on the contrary, that it must decline within a period of time, which, when measured against the life of families or nations, is remarkably short. To quote: 'If our consumption of coal continues to multiply for 110 years at the same rate as hitherto, the total amount of coal consumed in the interval will be 100,000m. tons. We now turn to compare this *imaginary* consumption of coal [with] the estimate of the available coal in Britain, viz. 83,000m. tons within a depth of 4,000 feet . . . we cannot but allow that — *rather more than a century of our present progress would exhaust our mines to the depth of 4,000 feet, or 1,500 feet*

61

deeper than our present deepest mine . . . I draw the conclusion
. . . that *we cannot long maintain our present rate of increase of
consumption.*'[2]

Jevons's deductions were, of course, perfectly correct. What
he called 'this imaginary consumption of coal' never did happen, as indeed his exploratory calculations had shown the
impossibility of its happening.

Jevons believed that there was no substitute for coal and that
'with our coal *power* must pass from us'. Yet he thought it 'just
possible . . . that some source of energy now unknown may be
detected'. If this happened, he concluded, 'such a discovery
would simply destroy our peculiar industrial supremacy'.

We know today – nearly a hundred years later, that such a
discovery was made, that petroleum has taken the place of coal
as the world's greatest source of primary fuel, and that it has
been one of the principal factors that have destroyed Britain's
'peculiar industrial supremacy'.

The sagacious Jevons may have failed to recognise the poten-
tialities of petroleum – to which he devotes barely half a page –
but he did not fail to see the immense weakening of Britain's
economy that would result from Britain becoming a large fuel
importer. He thus posed the question: 'Are we wise in allowing
the commerce of this country to rise beyond the point at which
it can be long maintained?'

Today, nearly a century later, the question might well be
asked again. That is to say, when discussing the long-term
demand for fuel we might consider not only how much we
should like to demand but also how much it would be wise to
need. It will be admitted that the current spate of studies in the
subject deals exclusively with the former question and never
with the latter.

In long-term forecasting elaboration is a vice and not a virtue. If
anyone should wish to draw the conclusion that all long-term
forecasting in this field is really illegitimate, I should have
considerable sympathy with such a view. Of course, what is to
be considered 'long-term' varies from subject to subject. The
demand for electricity, for instance, can be forecast legitimately
over longer periods ahead than the demand for coal. But even in
forecasting the demand for electricity many years ahead one

should be careful not to underrate the fact that any definite action based on such expectations is an act of faith.

Doubtful as I am about the legitimacy of long-term forecasting, I have no doubts about the legitimacy of 'exploratory calculations'. To take a number of assumed factors — rates of growth, estimates of total reserves, and such like — and to test simply 'where it would take us', may be a highly instructive exercise. Now, this is precisely what Jevons did a hundred years ago. But people are so avid for forecasts that they almost invariably fail to notice the difference between a forecast and an exploratory calculation.

5

'World fuel pattern in AD 2000', *The Times,* April 11, 1963

There is no end to the debate on fuel policy, but most of it suffers from being based on much too short a forward look — seven years to 1970 or, at the most, twelve years to 1975. Four facts make it necessary to look further ahead. First, the absolute inescapability of the need for fuel in a highly industrialised society — if fuel fails, everything fails. Second, the non-renewability of 94 per cent of all the commercial fuel and energy used in the world — coal, oil, and natural gas — with renewable energy, i.e., water power, contributing a mere 6 per cent : hence the importance of the relation between current use of fuel and ultimate reserves. Third, not merely the endlessness in time of the demand for fuel and energy but also its rapid cumulative expansion throughout the world : no end of this expansion is yet in sight. And fourth, the long gestation period in the fuel industries : it takes many years to get things going, and some actions — like closing a colliery — may be virtually irreversible.

These four facts give to fuel and energy a very peculiar position in mankind's economic life — a position which demands a higher degree of responsible good husbandry and foresight in this field than is demanded in any other. To the extent that our powers of accurate foresight are limited the need for good husbandry and conservation is reinforced. It is precisely because we do not know the future with any kind of precision that we must act with the utmost caution and the highest sense of responsibility when it comes to fuel.

For the United Kingdom, as for several countries on the Continent, the crucial fuel policy question of the day relates to the future size of the indigenous coal industry. The answer to this question, if it is to be rational, will be derived entirely from the view one takes about the pattern of world energy up to, say, the year 2000.

An exploration of the future is therefore inescapable, however great the uncertainties. How much fuel is likely to be needed? What kinds of fuel are likely to be available, and how much of each? Will there be a buyers' market (as now) or a sellers' market (as five years ago)?

As regards fuel demand for the world as a whole, we may extrapolate certain well-established rates of growth in average per capita consumption. In the past thirty-seven years, since 1926, world fuel consumption has increased more than threefold; if it does the same again in the next thirty-seven years world fuel consumption in AD 2000 may be around 18,000m. tons of coal equivalent (compared with 1,570m. tons in 1926 and 4,800m. tons now).[1] Simple extrapolations of this kind are, of course, extremely chancy, but if we aggregate the forecasts made by the various industrialised countries and take into account the ambitious plans for industrialisation and urbanisation pursued in the under-developed parts of the world, we find that 18,000m. tons of coal equivalent in the year 2000 would hardly be enough. However that may be, it is enough to baffle the imagination. Where is all this fuel going to come from?[2]

In recent years, oil has been the most dynamic and expansionist major supplier. Are there any likely limits to the expansion of oil? Authoritative studies from inside the oil industry suggest that the limit might be an annual output of about 4,000m. tons of coal equivalent to be reached around the year 2000. The limiting factors, it is suggested, are ultimate reserves — not simply the reserves now proven which, of course, would not even last until the year 2000. At 4,000m. tons of coal equivalent, oil would be contributing only 22 per cent to the estimated total requirements of 18,000m. tons, compared with 35 per cent at present. This is indeed a surprising conclusion.

Even if we added 50 per cent to the maximum annual output rate of oil, raising it to 6,000m. tons of coal equivalent, oil would be contributing barely the same percentage as now.

Natural gas reserves are generally thought to be proportionately much smaller than oil reserves, and the possibilities of harnessing additional water power at reasonable cost and in suitable locations are naturally restricted. It is difficult to imagine that these two primary fuels could maintain their share in the total. Apart from coal, there remains only nuclear energy.

President Kennedy recently asked the United States Atomic Energy Commission for a 'new and hard look at the role of nuclear power in our economy', and the commission's report is now available. The verdict is a hopeful one: 'Nuclear energy can and should make an important and, ultimately, a vital contribution toward meeting our long-term energy requirements.' But a reactor type suitable for large-scale application is not yet available; breeders, which can breed fissile material, are essential because naturally fissile uranium is so scarce that 'if this were our only potential source, the contribution to our total energy reserves would scarcely be worth the developmental cost'.

The American Atomic Energy Commission hope that increased support for research and development of breeders 'would make breeder reactors economically attractive by the 1980s. Assuming this result we estimate that by AD 2000 nuclear power would be assuming the total increase in electrical energy production, and . . . that nuclear installations might actually be generating approximately half the total electric energy in the country', i.e. about 23 per cent of total United States' fuel and power requirements. This contribution, if it could be attained throughout the world, would amount to 4,000m. tons of coal equivalent and would roughly equal the contribution from oil. It needs no emphasis that the chances of developing the contribution of nuclear energy in less than forty years to over thirteen times the present size of the contribution of hydro-electric energy are slight.

Even so, the world fuel equation for the year 2000 could not be solved without postulating something like a trebling of the world's coal output. Maybe the pattern would then look like that shown in Table 3.5.1.

TABLE 3.5.1

	Million tons of coal equivalent	%
Coal and lignite	6,000	33 to 43
Oil	4 to 6,000	28 to 33
Natural gas	1,000	6 to 7
Water power	1,000	6 to 7
Nuclear energy	2 to 4,000	14 to 22
Total	14 to 18,000	

All these estimates are not prophecies but merely exploratory calculations or feasibility studies. No doubt the future has many surprises in store for us. Something may turn up; on the other hand, it may not. But these exploratory calculations are not being made just as a *jeu d'esprit*. The United Kingdom is becoming ever more dependent on fuel imports, and western Europe as a whole is already by far the largest fuel importer in the world. If the world fuel equation can be solved only by postulating the kind of heroic expansion set down above it may not be an enviable position to be too dependent on fuels from abroad. Imported fuels may become very costly indeed.

This is the crux of the matter. To what extent is it wise to insure against the possibility of steeply rising fuel import costs — or even of an actual physical shortage — by maintaining one's own indigenous fuel production in good shape? The amount of coal that can be produced in western Europe in the 1980s and 1990s will be determined by what happens to the coal industry in the 1960s.

The present world glut of oil, with abnormally low prices, is largely due to the protection, by way of oil import controls, which the Americans are giving their indigenous fuel industries. They do it for the sake of economic and strategic security, believing that it would be irresponsible to gamble with fuel. It is strange that western Europe, immensely less well-endowed with natural resources, seems to be relatively unconcerned about the security of its fuel supplies and much more interested in cashing in on the temporary situation of low import prices — which are themselves dependent, among other factors, on the American import restrictions.

6

From 'Why fight for coal?' *Coal Quarterly*, September 1964[1]

How, then, can we assess the situation that is likely to develop up to the year 2000? Many attempts have been made to tackle this difficult problem by highly elaborate methods of detailed forecasting. The amount of coal that will be wanted in, say, 1985 will depend on a great multiplicity of factors — the total demand for fuels of all kinds and their respective prices. These prices, in turn will depend on costs of production and transport, and all this might be overshadowed by technological developments affecting such matters as fuel efficiency, convenience of use, progress in automation, etc. Costs of production, in turn, will depend on productivity, levels of wages, costs of materials, etc., and each of these items again depends on a large number of others. Now, it is possible to take each factor separately and try to guess at its future development. But what is the basis of such guesses? In the end, we have no other basis for predicting the future than the experience of the past: we 'extrapolate' past trends. But what is the past — the last three years, or ten years, or twenty, or fifty? An unanswerable question. To tell the truth, such 'extrapolations' are seldom more than the expression of a highly conservative and unimaginative 'mood' — that the trends of the recent past are likely to continue for ever. Thus, the whole process of detailed forecasting, with all its spurious elaboration, in the end amounts to little more than a roundabout way of asserting that 'everything will remain as before only more so'. If coal has been losing business in the recent past, it will go on losing trade; if oil has been getting more competitive since Suez, it will continue to do so in the future; if there have recently been discoveries of gas deposits in Holland, there will in future be more and more discoveries, etc., etc.

It is when we compare earlier forecasts of this type with the events that have actually come to pass that we realise just how futile this method really is. To build up an elaborate 'model' of the future can be of value for teaching purposes, because, if well constructed, the 'model' will show up all kinds of subtle inter-connections between economic factors, which could easily be overlooked without such a 'visual aid'. But the detailed quantitative estimates constituting the model are generally quite useless for taking decisions which involve a longer-term commitment.

And yet, such decisions have to be taken all the same. On what are they to be based?

The alternative method is one of concentrating attention solely on the main driving forces which are likely to shape the future situation in its totality. This is what the National Coal Board have to do — and have been doing — as the trustees of Britain's largest natural resource.

7

From Coal — the challenging future. Address to Combustion Engineering Association, London, December 4, 1969[1]

We only know the past, and that short moment which is the present, and the first job of people who have to reflect about the future is certainly to try and understand the past.

Since Britain is no longer self-sufficient in fuel, the canvas has to be, not simply Britain, but the world. Let us go a little way back and see what has happened in Britain. I have chosen 1952. Perhaps one reason for choosing 1952 is that in that year the Government, being worried about the future fuel supplies, set up a Committee. It was called the Ridley Committee, composed of the best people they could find from private industry, from Government, from the Civil Service, from the academic world, and from the fuel industries, to look into the future. They decided to look ten years ahead, from 1952 to 1962, and they wrote a very impressive report.*

In this report they gave an outline of the way they thought

things would be developing, not only in terms of total require-ments, but also in details of the specific fuels. Of course 1962 is long passed and we have had a chance of comparing the Ridley Report, prepared with a great deal of care, and what actually happened. You will find that the total requirement of the country was very well estimated but the breakdown of the details in regard to specific fuels was hopelessly out.

There is a lesson in this, a lesson which is very often over-looked today. People make a great number of forecasts but they do not ask themselves on what the alleged predictability is based and how they are entitled to write the figures down. There must be some fundamental justification, one must distinguish bet-ween the predictabilities and the unpredictabilities. Assuredly it is the job of the economist to predict the predictable but at the same time he should recognise that there are unpredictabilities which cannot be glossed over by merely presenting a figure. This is not always understood today. Normally somebody develops a pro-forma and every little box in the pro-forma has to be filled in. Some boxes can be filled in on the basis of real predictabilities but others are the wildest guesses which do not add anything to real knowledge. We then find that good infor-mation about the future is spoilt and contaminated by bad information being treated the same way.

The first thing that we understand about the past is that there has been a rapid expansion of total requirements. This applies to Europe and the world, as well as to the United Kingdom. Secondly, as far as the United Kingdom is concerned, there has been a very rapid decline in indigenous supplies. In 1952 the country was self-supporting in fuel. At present 43 per cent of all the fuel used in Britain is imported. In Western Europe the percentage is even higher and last year the Western European Community imported 62·2 per cent of its fuel supplies. So there is a tremendous dependence by Western Europe on territories outside, not only outside its geography, but also outside its political power and influence.

Taking the world as a whole, in 1952 the total fuel require-ment, expressed in coal equivalent, was 2,900m. tons. For the last calendar year, 1968, it was 6,200m. tons. These figures are so large that one cannot expect anyone to really grasp them, and it is worthwhile to reflect on what these figures mean. One is taught by anthropologists that the savage in the bush can count one, two, three, and everything above three is many. My researches in this subject suggest to me that the educated Western European can count up to about two or three thousand and up to that limit every figure really means something — he can savour it, as it were — he can really experience it — but above that limit he becomes increasingly vague. If on the radio or television somebody means to say 530m. but he reads it wrongly and says 53m. nobody notices it. On the other hand, if he makes a mistake within the range of up to three thousand he is immediately corrected. One of the difficulties in our subject is that the figures we are dealing with are so large that our real existential relationship to them is nil — we do not grasp them.

However, in the sixteen years from 1952 the total world requirement has risen by 40 per cent, from 2,900m. tons of coal equivalent to 6,200m. tons. In 1952 60 per cent of the 2,900m. tons, roughly 1,750m. tons, was in solid fuel including lignite, 28 per cent was liquid, 10 per cent natural gas, and 2 per cent hydro. These proportions have changed very considerably. In 1968 coal no longer represented 60 per cent but 37·5 per cent, liquid fuel 42·3 per cent, natural gas 18·1 per cent, and hydro has stayed the same, having risen in proportion with the total expansion. Although coal has fallen from 60 per cent to 37·5 per cent, the actual quantity has risen from 1,750m. tons to 2,320m. tons of coal equivalent.

Such have been the tendencies of the past, and there have been countless so-called forecasters whose total contribution to thought has been to 'extrapolate'. Now, of course, naive extrapolation never proves anything and simply suggests that the future will be the same as the past — only more so. We have had this, and we still have it, in most walks of life. There is a population increase and then somebody extrapolates this to the year 2000 or 3000 and predicts that there will be standing room only. There is a short-period decrease and he complains that the population is dying out. None of this is really of much use.

What about the future of fuel? We have now reviewed the past in a rough and ready way. Is the future going to be the same as the past, only more so, or what is it going to be? What is going to be the world situation in, say, 1980?

At the National Coal Board we have always thought that the

only possible method of getting an understanding of the future is to reflect on the predictables, that is to say, the totals. The totals are produced by some sort of great ground swell which is relatively predictable. Now, it is not illegitimate, for instance, to project total population figures into the future because there is a general law of growth. If the population grows then the number of mothers capable of having children grows − and so there is a real law of growth behind this. Quite different, for instance, from a growth of productivity. If I learn how to be more productive, this does not prepare the ground for an even greater productivity thereafter. I may in fact be exhausting possibilities of growth. Population rates of growth are relatively predictable and so, I suppose, are their general economic aspirations towards a higher standard of life. These totals are relatively predictable, and they will determine the total situation into which the various details will fit.

In 1962 we published a brochure called *Meeting Europe's Energy Requirements*. Reflecting about the year 2000 our conclusion was that everything seemed to point to a total world fuel requirement in that year of 18,000m. tons of coal equivalent.

I have already quoted two figures for the world's fuel use: 2,900m. tons in 1952, 6,200m. tons last year. For the year 2000, about 18,000m. tons of coal equivalent seemed plausible in 1962. Three years later we published another brochure entitled *An Energy Policy for Western Europe*. Things were going a bit faster than expected. We felt we had to produce a range in our forecast for the year 2000, and we made it 18,200m. to 22,000m. tons of coal equivalent. Quite recently we have had another look at this and we felt that these were underestimates, that one had to count on a still faster growth, and we called it 23,000m. tons.

Simultaneously, a very large American research organisation called 'Resources for the Future, Inc.'[2] ventured a guess and they called it 26,000m. tons. In other words, all the thinking that has gone into this matter over a long period of years suggests an order of magnitude which lies around 20,000m. tons. One feels that this is where the drift of things is likely to take us, to a trebling of the present requirements in the short span of thirty years.

It might be said that the year 2000 is too far ahead for

forecasting, but one has to think about it all the same because it is now that decisions have to be taken, for instance, about new power stations. There is a lead-in period, first of planning and then of construction, and one would expect the power station to be productive for thirty years. This takes us even beyond the year 2000. What did we think in 1962 about the year 1980, which has now moved quite near to us? In 1962 we estimated that by 1980 the total world fuel requirement would be 8,000m. tons but in 1966 we realised that this was a serious underestimate and we put it at 10,000m. tons. The recent American study already mentioned puts it at 11,200m. tons. A period of ten or eleven years is of course relatively more predictable than one of thirty years, and there is a consistency in these estimates but a tendency to underestimate.

Looking at the world one is struck by the fact that the picture is very uneven.[3] I think it was Cervantes who said that there are two kinds of families in the world — the have and the have-nots. I indulged in a little speculation and divided the world population in terms of the rich and the poor because I think that the laws of movement, the laws of growth, would be different for them, as the rich are already at a very high level. Using United Nations material released in 1966 I decided to call those populations 'rich' where the per capita fuel consumption was more than one ton of coal equivalent per annum, and I defined 'poor' those with less than this. In this way the world divided into about 1,000m. rich and 2,300m. poor. The rich used nearly 5,000m. tons of coal equivalent and the poor 700m. tons, in other words, the per capita consumption of the rich was $4\frac{1}{2}$ tons while that of the poor was one-third of a ton. In 1969, the per capita consumption in the USA is about 10 tons per annum and in Great Britain something like 5 tons.

We know that the population is growing and we also know that it is growing somewhat faster in the poor countries than in the rich. While many people talk about population control, nobody really has the faintest idea of how to do it. I applied an annual population growth rate of 1·25 per cent to the rich, and a higher rate, 2·5 per cent, to the poor. I also had to apply a growth rate to the per capita consumption because, after all, no one is satisfied. We hear continually that our growth rate is not sufficient, although we belong to the rich, and that we should have a rate of growth every year of 3 or 4 per cent. To be modest, I applied a per capita growth rate to the rich of 2·25 per cent and a rate of 4·5 per cent to the poor because they are still at such a low level, although they are making determined efforts to improve their lot. Applying these growth rates to population suggests a world population of between 6,000m. and 7,000m. in thirty years' time, a figure with which the demographers generally agree. Applying both the population and the standard of life growth rates to the rich and the poor yields a total fuel consumption of 23,000m. tons of coal equivalent in the year 2000. 'Resources for the Future' quoted 26,000m. tons and some years back we mentioned 18,000m. to 22,000m; so this figure is roughly in line.

One interesting point that emerged from this calculation was that by the year 2000 the average consumption by the rich would be at the level of the current consumption in the USA. But the poor would have grown to 5,000m. and would still, in terms of fuel consumption, be at the level of poverty – about 1·4 tons.

What is more important is to follow through such an exploratory calculation by making an aggregation. This indicates that in the thirty-four years from 1966 to the year 2000 the aggregate fuel consumption would be 425,000m. tons. It is interesting to compare this aggregate with the known world reserves as quoted by 'Resources for the Future', which state proved oil reserves of 94,000m. tons of coal equivalent and gas reserves of 50,000m. tons. The figure for coal reserves is of a different order of magnitude at 15,000,000m. tons, and these reserves are mainly in the USA, USSR, and China.

Now we know, of course, that one cannot proceed from proved reserves, but the moment we leave the solid basis of fact we are speculating. All I will say is that if we double the oil reserves from 94,000m. tons to 200,000m. tons of coal equivalent and double the gas reserves from 50,000m. tons to 100,000m., this still does not remotely fill the bill of 425,000m. So to the extent that mankind needs fossil fuels, the long-term future, as seen from now, would seem to lie with coal rather than with oil or gas. The Americans are very much aware of this and they are spending a lot of money on the gasification of coal, and on turning coal into a liquid.

73

8

From Facing the change in the energy situation. Conference organised by the Royal Institute of British Architects, University of Durham, July 18, 1974[1]

Until a few years ago, we took oil for granted. In the autumn of 1967, for instance, the then Minister of Power presented a White Paper to Parliament, entitled 'Fuel Policy'.[2] Its central message was this:

> Subject to overriding considerations of adequacy and security of supplies, the Government's basic objective can be summarised as cheap energy . . . What is important is that we should take full advantage of the cheapness and technical merit of nuclear power, North Sea gas, and oil.

The Minister had no real doubts about the adequacy and security of oil supplies and therefore mapped out the further contraction of the British coal industry, the speed of which, however, would have to be controlled so as to avoid undue hardship to coal miners and their communities. In a sagacious paragraph (Para. 53) he dealt with future oil costs.

> It is difficult to predict the course of oil prices. There are a number of reasons for expecting them not to increase. The industry is continually searching for ways of cutting costs, as for instance by the use of very large crude oil tankers to reduce freight charges and increase flexibility and security of supply, and the surplus of crude oil seems likely to persist for many years despite the expansion in world demand. Here and elsewhere oil will be up against increasingly strong competition from natural gas and nuclear power. On the evidence available, it seems likely that oil will remain competitive with coal, and that pressure to force up crude oil prices will be held in check by the danger of loss of markets.

These sanguine arguments did not go unchallenged, but no one was willing to listen to the challengers, because the enticing dream of 'cheap fuel for ever' kept everybody happy. We now have a very rude awakening. The following official figures make the point: in 1970, our average monthly crude oil imports amounted to 8·3m. tons; at an average value per ton (f.o.b.) of £4.80, these monthly imports cost the country £40m. In March 1974, our crude oil imports were rather high at 9·6m. tons; the average value per ton (f.o.b.) was £28.09, and the total bill came to £278m.; compared with four years ago, an increased burden on our balance of payments of nearly £240m. *a month*.

'Pressure to force up crude oil prices' (we have been told) 'will be held in check by the danger of loss of markets.' Do you think the oil exporting countries are worried about a loss of markets when the export of one ton of crude oil produces for them an amount of money which, only four years ago, required the export of six tons? On the contrary, they would rejoice in the loss of markets, if only they could see it happening. They have been pleading for years that the oil importing countries should *reduce* their requirements. From their point of view, a loss of markets is the very thing they have been longing for ever since they realised — some ten years ago — that their proved oil reserves were by no means infinite and would be exhausted in a matter of twenty or thirty years if they 'maintained their markets'. In ever more insistent tones they have pleaded: 'Please mitigate your requirements; if we sell all we have got within twenty or thirty years, *what is to become of us*? We cannot build up an alternative livelihood for our peoples within two or three decades.'

4

The Causes of Crisis
Shutting Down the Mines

Introduction

In the years after the war, which coincided with the early years of coal nationalisation, energy was chronically scarce in Britain and the rest of Europe: it was the factor most restricting economic recovery. The badly under-capitalised coal industry found expansion difficult and slow. The NCB put together its first investment plan aimed at raising output to 240–250m. tons in 1965.

Inland coal use increased from 184·5m. tons in 1947 to 217·5m. in 1956. Exports, 5·3m. tons in 1947, rose to 19·2m. in 1949, but were down to 9·7m. by 1956 because the country had insufficient to spare. In fact, coal had sometimes to be imported at a much higher price in this period.

The industry in 1957 was shaping itself to meet further growth in demand when cheap and plentiful oil started to flow in. The rest of Schumacher's time at the Coal Board was dominated by attempts to resist contraction of the mining industry caused by this competition, and later, by nuclear power.

During 1958 and 1959, 85 collieries employing nearly 30,000 men were closed: almost all of them were nearing the end of their useful lives and nearly all the men who wanted to stay in the industry were able to move to more modern, productive mines, though in some cases these were 200 miles away.

Successive British governments were convinced that the only reasons why the industry could not be cut back even faster were the social consequences, but the West European coal producers continued their consistent campaign for a viable industry.

76

Schumacher helped to draft three reports published by the Association for Coal in Europe.[1] In the 1960s the Coal Board under Lord Robens, whose determination and political skill were backed by Schumacher's clear analysis of the long-term needs, fought and argued publicly against Conservative and Labour governments.

Despite a partial recovery between 1962 and 1965 caused by big improvements in operational efficiency and some government aid, the long-term decline of the industry continued. Output went down from 207m. tons in 1956 to 133m. in 1971, the number of collieries shrank from 840 to 292, and the man-power from nearly 700,000 to less than 290,000. Throughout some years an average of more than one mine a week closed.

The industry's confidence suffered perhaps the biggest shock of all in November 1967 with the publication of the Labour Government's Fuel Policy White Paper.[2] Despite the war in the Middle East only a few months earlier, the Government based its policy on 'the expectation that regular supplies of oil at competitive prices will continue to be available' and the assumption that 'on any tenable view of the longer-term pattern of energy supplies and costs, the demand for coal will continue to decline'. Yet only two months before the White Paper was published Prime Minister Wilson asked for the deferment until 1968 of the closure of sixteen collieries, planned for the last three months of 1967, because of expected high unemployment during the winter.

It quickly became clear that the government-aided and government-inspired cut-back had been too drastic: in the winter of 1970–1, undistributed coal stocks virtually disappeared; coal had to be imported at prices far higher than those of indigenous coal; and oil prices soared in response to a possible fuel shortage. The two new fuels, nuclear power and gas from the North Sea (which had encouraged the Government to talk of a four-fuel energy economy), together contributed less than 8 per cent of the country's total needs in 1970.

Gas, first proved in the North Sea in 1965, made rapid progress though, and only two years later was contributing the equivalent of 36·4m. tons of coal. Schumacher was among the first of many people who argued – unsuccessfully – that a resource of such comparatively short life should be depleted at a controlled rate, and warned of the effect a crash programme for its introduction would have on the coal industry. North Sea oil did not come into the reckoning until later: the first field was proved in 1969 and it was 1976 before sizeable supplies began to come ashore.

Again, the coal industry started to plan for expansion. However, it was not until after the 1973 Arab/Israeli War, with national economies stricken by the world-wide crisis that followed, and after two strikes in the coal industry, which were partly a reaction to the long crisis, that the Labour Government started to provide the resources the Board wanted to invest in coal output — despite the fact that oil and gas from the North Sea would for a time help to make Britain self-sufficient in energy supplies. The coal industry's exploration work has been extremely successful: new reserves are being proved at a rate four times as great as the present production and use. Governments and the country appreciate that coal reserves sufficient for about 300 years are a national asset of great value, representing a potential source of economic and political independence.

Though competition from oil and later from natural gas were the main problems during Schumacher's time at the Coal Board, business was lost also to nuclear power, but for non-commercial reasons. The official justifications for the first phase of its development when nine Magnox reactors were built as a result of government decisions in 1955 and 1957, were that Britain must be involved in the development of this new energy source and that coal would not be able to meet the whole of the growing power station demand.

The Coal Board did not oppose the case but they did question, given the experimental nature of the programme, whether it was necessary to build so many stations. Robens pointed out that the nuclear stations built by 1970 were already displacing nearly 10m. tons of coal a year, representing the output of almost 18,000 miners.[3] This business was not lost because coal was uncompetitive in price: if the nuclear stations had been charged their share of development costs (actually paid for out of public funds) they would have been hopelessly uncompetitive.

The main clash with nuclear energy, however, was over the second power station building programme based on the

78

Advanced Gas-cooled Reactor (AGR), with the first order placed in 1965.

Eventually, the Coal Board and their allies in Parliament and elsewhere were justified in their opposition. Although some of the stations have performed well, the AGR programme as a whole has far from fulfilled expectations: in fact Duncan Burn[4] describes the programme as a disaster. The repeated error of building five big stations without sufficient experience of commercial-scale operation, specifically warned against by the NCB, has cost the electricity consumers very dearly indeed. No AGR station has been exported. But the programme cost the coal industry more business and many thousand miners their jobs.

1

From Long-term demands for fuel. Paper to a study group of the Royal Statistical Society, May 21, 1958[1]

Every 100 minutes the earth receives as much energy from the sun as mankind uses in a year. Man is not interested in energy as such, but only in *low-cost* energy.

2

From Investment in coal. Address to National Coal Board Summer School, Oxford, September 1955

We know that we have much to do and that there is much room for improvement. But we also know that we are up against a job of altogether unusual difficulty to which there is no quick and ready solution. Few people outside the industry have any comprehension of it. Why is this so? Because most people cannot grasp the fundamental difference between ordinary manufacturing enterprise on the one hand and extractive industry on the other. They do not see that an old extractive industry unavoidably loses capacity in the ordinary course of its operations all the time; that it always must engage in a great deal of new exploration, new investment and new development — not to expand its output capacity but merely to maintain it or stop it falling.

Nor do they understand the time involved in doing anything new in coal. You can build a new factory or refinery in two or three years. But you cannot develop a new colliery, or reconstruct an old one in that period. Why not? Because the main structure of a colliery consists of shafts and tunnels through stone. In construction work on the surface you can deploy hundreds and even thousands of men all at the same time; you can build all the different main sections of the project simultaneously. But in the opening up of a new colliery you cannot do this; you cannot start tunnelling underground before you have a shaft to get underground. You cannot tackle the main sections of the job simultaneously, but only one after the other. And what is involved in sinking a shaft, for instance? Can you deploy thousands of men to sink a shaft in record time? Of course, you cannot. You have only a few square yards, only the size of a squash court, on which to deploy your men. And the same with tunnelling.

During most years of the inter-war period the British coal industry was shrinking owing to a decline in the demand (mainly the export demand) for British coal. So the inevitable shrinking of capacity passed unnoticed. Capacity fell, but it remained always just a bit larger than that required for meeting the demand for coal. For us, 'it takes all the running we can do to remain in the same place'. Our predecessors did not have to run because they were not required to remain in the same place.

What ought to have come to our help during the last few years was the completion of large-scale investment projects started before or during the Second World War. But for obvious reasons very little was in fact started then, and very little has therefore been coming forward.

The whole industry is involved in a tremendous struggle to maintain and ultimately to enlarge its output capacity. It is estimated that the annual loss of productive capacity, arising simply out of the current operations of this extractive industry, amounts to as much as 4 or 5m. tons. This means that in the absence of large-scale reconstructions and the opening up of new pits, and in the absence of measures to increase the rate of utilisation of existing capacity, the industry's output could be expected to fall by 4 or 5m. tons every year. I suggest that the industry can be proud of itself because its annual output, in fact, has risen by some 30m. tons since Vesting Day.[1] Output has been increased in spite of a fall in capacity, and this is due primarily to a very large number of improvements introduced in practically every pit of the country.

Why then should there be this feeling of crisis? Because the country's total demand for fuel is rapidly and continuously rising while consumption of coal — the source of 90 per cent of all fuel supplies — is outstripping production. And this is mainly the direct result of the country's industrial prosperity.

It is quite true that there is a threat of crisis. But it is the threat of a fuel crisis for Britain, not of a crisis in the coal industry. Because British agriculture cannot fulfil all home demands, would you therefore speak of an agricultural crisis? That would be a gross misuse of terms. There is a shortage of coal also on the Continent, but there is no talk of crisis in the Continental coal industry. Steel is short both in Britain and on the Continent, but there is no crisis in the steel industry.[2]

I should say to you therefore that we should not be depressed by this talk of crisis, that we should not fall for it. The coal industry, with all its shortcomings, is healthy and enterprising. It is not collapsing or breaking down or going to pieces. More constructive and enterprising work is going on in the industry today than has ever been seen before. It is not a trivial matter, nor one easily accomplished, that the Board have now for some years been doing as much exploratory boring in twelve months as was done by our predecessors in twenty years. A hundred or so large reconstruction schemes in the process of execution up and down the country is not a sign of a lack of enterprise, vigour, or daring. And it imposes a burden of work hard to appreciate by outsiders.

At the recent Atomic Energy Conference at Geneva it was suggested that, by the end of the century, Britain would be using between $2\frac{1}{2}$ and $3\frac{1}{2}$ times as much energy as she does now. It was also suggested that atomic energy might by then contribute half of the required energy supplies. This would still leave the other half for the conventional fuels. One half of a total that has grown to $2\frac{1}{2}$ or $3\frac{1}{2}$ times its present size is between 25 and 75 per cent more than the present total. In other words, if you look at the story as a whole, and not only at half the story you find that the Geneva picture shows up a much larger need for the conventional fuels in the year 2000 than the present need.

The two sides of this picture belong together. You cannot separate them. If there is no large and sustained increase in national income and productivity during the next forty-five years — and therefore no large and sustained increase in total energy requirements — then the material resources for the assumed enormous development of atomic energy will be lacking. It is only when the total economy maintains a high rate of annual growth that large-scale atomic energy will be both necessary and possible. And that general growth itself is a guarantee for further increases in the demand for coal and oil.

It is a pity that people spend their time in coining smart phrases like 'the atomic writing is on the colliery walls'.[3] This is like scuttling your lifeboat when you have sighted an unknown island fifty miles away. No one yet knows, nor shall we know for many years, whether the island is inhabitable. And even if it is, it is quite certain that we shall need the lifeboat for numerous other purposes.

3

From What should be our national fuel policy? Notes for an informal meeting, Institution of Electrical Engineers, London, February 6, 1956

Can we simply neglect the fact that the country would be making itself dependent so very largely on *imported* oil? And where is all this oil going to come from? Almost exclusively from the Middle East — politically the most unstable part of the world.

This, in my opinion, is the crux of the matter. If we are happy about making ourselves dependent — along with the rest of Western Europe — on oil supplies from the Middle East, if we are prepared to run the inevitable risks of being dependent on no major disturbances occurring in this disturbed part of the world, we can let fuel requirements 'rip', as depicted in the figures I have presented. Technically, I have no doubt, this can be done. But is it wise to do it?

4

From 'The new energy pattern in Europe:[1] an analysis of the Robinson Report', *The Manager,* Volume 28, September 1960

The new report . . . is not much troubled by any 'inherent risks' of Western Europe's increasing dependence on imported energy, nor worried by a 'certainty of increasing prices'. 'When formulating a long-term energy policy', says conclusion (v) of the new report, 'the paramount consideration should, in our view, be a plentiful supply of low-cost energy with a freedom of choice to the consumer.'

Cheapness, not any uncertainties about continuity of imported supplies, the balance of payments, future prices, etc., should be the 'paramount consideration'. 'We recognise', says conclusion (vi), 'the importance of continuity and regularity of energy supplies. But we do not regard the long-term protection or artificial encouragement of indigenous supplies of energy as the most satisfactory method of obtaining such security.'

It appears to take it for granted that any future change in the relativity of fuel prices could only be to the disadvantage of coal. As a result, the Robinson Commission recommends the abandonment of whole 'regions' of coalfields and the concentration of European coal production solely on the best resources. The Commission even takes it upon itself to warn governments against delay: 'Where gradual and progressive concentration of production is practicable, we hope that it will be adopted. But we doubt the wisdom of greatly delaying the necessary process of adjustment' (paragraph 194).

It is of course unwise to delay a painful adjustment that is known to be necessary and inescapable. But if the necessity and even the desirability of a given course of action are open to doubt, delay is the only reasonable policy. It is doubtful whether it would be wise for Western Europe to abandon and thereby largely to destroy, a substantial part of its indigenous fuel base. The Robinson Commission believes that 'in the long run' (paragraph 194) a smaller coal industry would be preferable, in spite of rapidly increasing fuel requirements. It is not disturbed by the prospect of an immense increase in Western Europe's dependence on imported oil. Yet in another context, when defending 'an active policy of developing nuclear energy' (which according to paragraph 153 'secures a relatively small saving of foreign exchange at the cost of a very large investment of indigenous European resources') the Commission itself points to: 'a later stage when the growing world demands for energy begin to exhaust the world's supplies of oil' (paragraph 153).

The 'later stage', invoked as a justification for very large expenditure on nuclear energy now, is evidently not considered to be very far off.

It is the central weakness of the Robinson Report that it does not look beyond the year 1975, fifteen years from now. There are only a few references to the longer term. One of them occurs in paragraph 136: 'While the need to develop new sources of energy is a real one in terms of a much longer period than we have under review, it does not seem likely that shortages of oil supplies will make themselves felt in an acute form by 1975.'

In short, the Commission has reviewed only the period up to 1975, and all it can promise is that shortages of oil supplies in an acute form are not likely to make themselves felt by then. The Commission leaves unclear how it arrives at 'the essentials of a long-term energy policy' (paragraphs 170–176) and at its recommendation to curtail and concentrate the coal industry in the long run. Its policy recommendations, if followed, would lead to practically irreversible actions of the greatest import for the period beyond 1975. It has not reviewed that period, and its only references to the longer run are warnings about possible shortages of oil supplies. The Robinson Report thus fits the description which Mr George Kennan has recently given of certain tendencies in the public philosophy of the West — a self-centredness that regards 'the convenience of contemporary man as an end in itself, as though there were no past and no future'.

84

The international oil companies claim to earn considerable amounts of foreign exchange for Britain; but it does not follow that therefore the expenditure of foreign exchange on oil imports constitutes no burden to the balance of payments. All commercial imports cost foreign exchange; whether that foreign exchange has been earned by the importer himself or by someone else makes no difference. If there were no 'net' exporters, there would be no foreign exchange to pay for Britain's food imports. A further doubling or even trebling of energy imports, mainly in the form of oil and natural gas, as envisaged by the Robinson Commission, would impose severe strain on the balance of payments of the United Kingdom. Strains have already been noticeable during recent years, while the value of oil imports (net, that is, after deduction of the value of oil exports) has been climbing up to nearly 10 per cent of all United Kingdom imports, thus constituting the biggest single item in Britain's import bill.

The keynote of the Robinson Report is the plea for laissez-faire in the fuel economy of Western Europe; coupled with the advice that prices should 'fully reflect the costs and scarcities of individual types of energy' (conclusion xi). It is held that the general interest is best served 'if all possible steps are taken to ensure that competition between different sources of energy is fair and that distortions are either eliminated or adequately compensated' (conclusion v). Questions relating to the problems of balance of payments, security of supplies, and the like are dismissed. The Report thus reflects the current climate of opinion in Western Europe which generally favours free trade and reliance on the price and market mechanism. If the Report dealt with ordinary goods and services, such as constitute the major part of all economic exchanges, it might deserve nothing but praise as an informative and well-balanced piece of economic analysis and forecasting.

But the Report deals with fuel and energy, which are different from the common run of economic goods and services in that they are mainly derived from non-renewable sources. When dealing with non-renewable resources, there is need for the most careful consideration on the problem of conservation. There is also need for looking much further ahead than would normally be required.

5

From The coal industry and its contribution to Britain's energy requirements. Address to Midlands meeting of Combustion Engineering Association, Birmingham, March 20, 1963

To close a pit is an irreversible decision. Once a pit is closed it is not possible to re-open it except by the expenditure of a vast amount of capital. To keep a pit on a care and maintenance basis is so expensive that, in fact, it is never done. So these irreversible decisions have to be taken and they have to make sense, not just now, but also twenty, thirty or forty years hence.

6

From The struggle for a European energy policy. Paper at the International Coal Conference of the Japanese Coal Association, Tokyo, October 1963

It is not to be expected that a 'European Energy Policy', freely adopted by all European countries, will soon emerge. Countries without indigenous resources may continue to favour a policy of 'bargain hunting' in the world's energy markets, and the burden of conserving Europe's indigenous fuel resources may continue to fall exclusively on those countries in which these resources happen to be situated. It is not surprising that business men in the latter countries should be inclined to look enviously at their competitors elsewhere who can reap the fullest benefit from the present fuel 'buyers' market'. This situation, however, will change dramatically, as soon as the first signs of rising prices and tighter markets appear, and the meaning of the words 'European Energy Policy' will also change. It will then be held that Europe's coal must be fairly distributed to all who need it, that it must not be reserved to users in the country of its origin. The trouble is only that then it might be too late. The European coal producers rightly insist that:

a contraction of the coal industry today and during the

coming years would be virtually irreversible. The issues at stake are therefore very great indeed, extending far beyond the prosperity of the coal industry itself and involving the prosperity and security of the whole Western European economy.[1]

7

From 'A new look for coal', *Daily Telegraph,* January 20, 1964

Everything begins with people, and the greatest single bottleneck in the coal industry during the first ten years or so after Vesting Day (January 1, 1947) was the shortage of highly-trained manpower on the side of management. The 'pipeline of education', as everybody knows, is a long one in years.

So is the 'pipeline of investment' in coal. You spend money today but you cannot hope to see results for perhaps eight or ten years. This is hard to bear when everybody is desperately pressing for higher output (as was the case until 1957) and better performance. New machines, too, are not developed overnight, and it takes many years, even after a technical breakthrough, until a sufficiently wide range of new machines has been developed to suit the great variety of British seam conditions.

This is past history. The efforts of the past are bearing fruit today. So this is not a flash in the pan; it is the outcome and reward of many years of very hard work.

Yet there is more to it than that. Under the leadership of Lord Robens a new psychological climate has been created in the coal industry which (to stay in the metaphor) makes all fruit ripen faster and better than ever before. Nothing succeeds like success, and the 'image' of success is itself conducive to further progress.

Let no one underestimate these psychological factors. They weigh more heavily in the coal mining industry than almost anywhere else. Why? Because to go into mining as a miner or a mining engineer means to make yourself, so to say, a captive worker.

A mechanical or electrical engineer can find a livelihood and a rewarding career, according to his abilities, in a very wide range of industries; but a mining engineer commits himself well-nigh irrevocably to the future of mining. If, therefore, there is a pervading suggestion that coal mining is 'on the way out', who will want to enter such a commitment?

One only has to look across the Channel to see how coal industries can be 'talked to death', or nearly so. The mining departments of some of the world's finest technical universities are almost empty of students, except for foreigners. The recruitment of indigenous miners dries up; wastage increases; and replacements can be found only by combing foreign lands, from North Africa to Turkey and even Japan.

What had started as an unjustifiable doubt about the future need for coal quickly becomes a justifiable doubt about the coal industry's ability to supply even the undoubted requirements of the market. Without well-qualified men, nothing can be done.

Some two or three years ago, the British coal industry was equally in danger of being 'talked to death'. But this danger, happily, has passed, and everybody in the industry knows whom to thank for it.

It is hard, and perhaps unnecessary, to single out one particular action or event that turned the tide. In my opinion, it was Lord Robens's courageous announcement, nearly three years ago, that the coal industry will sell 200m. tons of coal a year, and that it will be kept in a fit condition to produce that amount of coal every year for many years to come. There is a growing need for fuel and energy all over the world, and the industry's task is to conquer and hold a market for 200m. tons — that was the message.

For the first time in many years, this established at least one firm point in the sea of general uncertainty. 'How can Robens stick his neck out like this?' people asked. At that time, total sales were falling below 190m. tons. Well, in 1963 the industry sold 201m. tons.

'How can you be sure that you can hold the position?' people ask. The message went out that the target should now be raised to 220m. tons a year. Why 220m. tons? Because there is unused capacity in many of Britain's best pits, and the next 20m. tons a year, won from these pits, would be the cheapest fuel Britain

88

could buy anywhere in the world.

This has come to be called 'marginal tonnage'. It could be cheaply got, because the pits are there, fully equipped and fully manned 'elsewhere underground' and on the surface: to get extra coal from a few additional coalfaces would require only a small number of additional faceworkers. If a market can be found for that extra coal, all coal will be cheaper.

New possibilities for labour-saving mechanisation are emerging with regard to almost every operation underground and on the surface — developments in powered supports at the face, the roadhead and the roadway; stable hole and heading-out machines; ripping machines; cable-handling devices, and so on. Promising work is going ahead on gas control and firedamp drainage; on automatic monitoring systems of roof supports; on instrumental aids for management to increase the rates of face advance and to promote the automation of underground transport; on instantaneous ash-monitoring devices and many other matters relating to coal preparation.

Coal has been mined in Britain for many centuries; yet today, strange to say, we have a new coal industry.

8
From 'Why fight for coal?', *Coal Quarterly,* September 1964[1]

The British coal industry is fighting, and fighting hard. It is struggling to hold a market of at least 200m. tons of coal a year. This requires a tremendous effort along the industry's entire front line which extends from the coalface right to the point at which the coal is turned into useful heat or energy by the coal consumer.

It is fighting against heavy odds. Coal's main competitor, fuel oil, is a 'joint product' without clearly definable costs of production. Fuel oil is generally sold at prices below those paid for the raw material — crude oil — from which it is made. How is it possible to offer a refined product at a lower price per ton than the price (per ton) of the crude material? Simply because high profits can be made from some of the other 'joint products' of refinery operations, such as petrol and lubricants, which are monopoly products in the sense that there is no substitute for them. Coal is not a 'joint product': it cannot push its production costs over to some other article, as is the case with fuel oil.

Another competitor is nuclear power, which no one even expects to compete on economic terms. All the same, it does compete and it absorbs a part of coal's natural market, supplying heat for electricity generation at a cost which can be justified only as 'development expenditure'.

It would be easy indeed to give up and lower the target. Why fight for the marketing of the 'last' 20 or 30m. tons a year? Why not close down another 100 collieries and be content with the cheapest coal from the best pits only?

It is the National Coal Board's duty, imposed by Act of Parliament, to 'further the public interest in all respects'. As no one has yet defined authoritatively 'the public interest' — which is in any case a many-sided thing, not easily definable and never stable in time — it is in fact for the Board themselves to define it. Why not go into retrenchment? Why not yield to the pressure of competition, whether that competition be 'economic' or not?

The Board obtain loyal service from their employees who deserve loyal treatment from the Board. Even if it were true — which it is not — that the country would be better off with a smaller coal industry, any attempt to close a great number of collieries all at once would throw the whole industry into turmoil. Thousands of men would have to be dismissed precisely in those parts of the country where there is already higher-than-average unemployment, and the Board would totally lose the confidence and loyal co-operation of the men. What would then be the cost of producing the coal without which the country simply cannot function? These are 'social considerations', and too many people still argue as if a business like the coal industry could operate more cheaply or more efficiently if it paid less scrupulous regard to its social obligations and responsibilities. The truth is that it is utterly uneconomic to neglect them. If anybody cannot see that the humane treatment of an industry's labour force is a good thing in itself, he should at least be able to understand that the opposite is extremely costly to all concerned.

90

There is, of course, free consumers' choice. But the consumer, no less than the Coal Board, has to consider the longer term as well as the short. There can be no question of indigenous coal having to withdraw 'from some of its less sophisticated uses' such as burning it under boilers, in the 1980s or later. We have been warned that this possibility could arise with another fuel, which is not indigenous. It is impossible to put a money value to 'security of supply': that it has a value all the same, which may mean very big money at some future date within the foreseeable future, could hardly be open to doubt.

9

From Coal in the fuel economy. Paper to the Fourth International Mining Congress, London, July 15, 1965[1]

A continuation of the rate of growth experienced during the last twenty years would carry world fuel needs to the staggering total of about 20,000m. tons of coal equivalent by the year 2000 — in thirty-five years time.

The question of whether the world will be able to cover its fuel requirements without drawing heavily on coal must sound a strange one to anyone who has comprehended these magnitudes. If total supplies have to double in fifteen years or so, is it likely that this can be accomplished while neglecting coal, which at present accounts for about one-half of total supplies? What is there to take the place of coal and at the same time to look after the requirements of growth?

The first suggestion is that nuclear energy will totally change the picture. In Britain, considerable excitement has recently been caused by the announcement that a new type of nuclear reactor is now on the drawing board which promises to be capable of producing electricity at a lower total cost per unit than coal or oil fired stations of the latest design. It is called the Advanced Gas-cooled Reactor, a near-prototype of which — the Atomic Energy Authority's 35 Mw reactor at Windscale — has been operating satisfactorily for something like three years. The step-up from a 35 Mw prototype to a 1,200 Mw commercial station, of course, creates a number of uncertainties; but being unable to pronounce on the reliability of the promise that unit costs will in the event be as low as now stated, I accept it at its face value.

How will this affect the world fuel situation? Electricity derived from nuclear reactors involves a very substantial increase in capital costs associated with a great reduction in running costs; it also involves production in very large units. This means that it is a process available only to countries rich in capital resources, where there are large concentrations of population and industry. It means, furthermore, that nuclear stations are likely to be economic only when used for supplying base load.

If the first of these stations comes on stream in 1970, five years from now, what is the quantitative impact of this new source of primary fuel likely to be ten years later? I am speaking of the world as a whole, and in the world of 1980, as we have seen, total primary fuel requirements are likely to approach 10,000m. tons of coal equivalent. It is obvious that a few dozen such stations, each with an annual coal equivalent of, say, 3m. tons, would have no significant impact on the world fuel situation at all. Over thirty stations would be needed to account for about 1 per cent of the total world fuel requirements; over 300 stations, to account for 10 per cent. How many will, in fact, be built in ten years?

It would seem safe to assume that there will be a few stations outside Europe, Japan, and the United States. The latest forecasts made in these countries suggest that the total installed capacity of nuclear power in 1980 would be of the order of 120,000 Mw or probably less than 300m. tons of coal equivalent.[2] While these are very impressive figures which, if realised, would represent a truly colossal achievement, it is worth remembering that 300m. tons of coal equivalent represents only about one year's growth in the world's total requirements. Now it is certainly welcome that this new source of primary fuel may be able to take care of the increment of requirements, statistically speaking, of one year out of the next fifteen, and the question may well be asked why the programmes might not be doubled or trebled so as to take care of two or three years' growth out of fifteen. There are no doubt quite a few reasons which have induced the experts, in spite of their general optimism, not to put forward such very much higher figures. One reason may have been a realistic appreciation of the enormous capital costs involved. Another may be the weighty doubts that still attach to the safe and economical disposal of radio-active wastes. It is also somewhat uncertain what is to become of the nuclear reactors themselves after twenty or thirty years when they have ceased to be productive yet continue to be highly radio-active. What will be involved in the task of making them safe for centuries to come, and how can one comfortably contemplate the accumulation of an ever greater number of abandoned centres of radio-activity, particularly in the most densely populated countries of the world? The main reason, however, why the experts do not hold out the hope of a much greater expansion of nuclear energy during the next fifteen years — in any case, the reason freely discussed in public — is an expected shortage in the supplies of uranium. In the Euratom report, *The Problem of Uranium Resources and the Long*-term Supply Position, a computation of uranium requirements in relation to total reserves discovered to date shows that the supply position is in no way assured. In the absence of substantial new discoveries, the reserves by 1980 would be reduced to three or four years' consumption at the assumed 1980 rate. Although there are large reserves of low-grade ores, the Euratom study does not regard them as a feasible source of supply. It takes about ten years from the discovery of new reserves to the start of production.

For all these reasons it would appear to be rash to disagree with the experts when they warn us that there are strict limitations to the expansion of nuclear energy as long as nuclear reactors, whether based on natural or enriched uranium, are incapable of utilising the 99·3 per cent of the element which consists of isotopes not naturally fissile. The verdict given by the American Atomic Energy Authority[3] in November 1962, still seems to stand and is worth quoting:

The fissionable material found in nature is confined to uranium-235, constituting only 0·7 per cent of natural uranium . . . If this were our only potential source, the contribution to our total energy reserves would scarcely be worth the developmental cost. (p. 22)

This means that a really massive contribution from nuclear energy to the world's fuel requirements will depend on the development of economic breeder reactors, and

> with luck and adequate effort, practical and economic full-scale breeder reactors might be achieved by the late 1970s or early 1980s. (p. 51)

But even then certain narrow limitations on growth will persist:

> Even when breeder reactors become economic and begin to be installed there will be a complication regarding fuel supplies. At least for some time to come, economic breeders will have breeding gains so low that they will produce not more than 3 per cent or 4 per cent of their fuel inventory each year . . . The time required for a reactor to produce enough excess material to fuel a second reactor . . . will probably be 15 to 20 years, or even longer for the first economic breeders. (p. 39)

If the position is as thus described by the experts, two conclusions would appear to be reasonably safe; first, that the appearance of an economically viable Advanced Gas-cooled Reactor in 1970, important as it will undoubtedly be, will not remove the limiting factors which arise from the relative scarcity of high-grade uranium resources, and second, that the contribution of nuclear energy *by 1980* is unlikely to exceed something of the order of 3 per cent of world fuel requirements,[4] a contribution well within the inevitable margin of error of all demand estimates fifteen years ahead. The world economy in 1980 will therefore be based on fossil fuels in much the same way as the world economy of 1965. What may happen after 1980 is not my present concern.

There is a good deal of talk about the diversification of oil supply sources, which, of course, is a matter of very great importance. To find over 60 per cent of the world's proved oil reserves concentrated in the somewhat unstable countries of the Middle East is hardly a comfortable situation. Yet diversification away from the Middle East appears to be extraordinarily

94

difficult. The Middle East accounted for 61 per cent of world oil reserves in 1954; again for 61 per cent in 1958; and again for 61 per cent in 1964.[5]

The oil industry is, of course, highly conscious of the need to find new reserves, and to find them in places other than the Middle East. Hence their willingness to spend enormous amounts of capital on exploration in the North Sea, in spite of the fact that every borehole there is several times as expensive as a borehole on land. In the interest of humanity as a whole we must wish them luck, because unless they are highly successful pretty soon, that is, within the next ten years or so, world-wide economic expansion will run into severe difficulties, not to mention the political tensions that would emerge.

The truth, as I see it, is that the expected doubling of world fuel requirements in the next fifteen years or so presents a tremendous task to all fuel producers. They will have to be very successful to meet it.[6] Fifteen years is not a long time when it comes to developing new oilfields, new coal-getting capacities, new uranium deposits, and so forth. There is need, therefore, for very deliberate policies of conservation and preparation for expansion.

The world will undoubtedly need massive contributions from nuclear energy as soon as they can be obtained at reasonable cost. But here again caution would seem to be necessary. If the development of nuclear energy, as is at present the case in Western Europe and particularly in the United Kingdom, merely forces the indigenous coal industry into contraction, there is no net gain from the whole venture in terms of fuel supplies. For every ton of coal equivalent supplied by nuclear energy a ton of coal-getting capacity is abandoned and, with it, a great many tons of coal reserves. The original intention of Britain's expensive nuclear programme was to create a source of primary fuel that would reduce Britain's dependence on imported oil. This intention, however, seems to have been forgotten, and the creation of nuclear capacity at immense capital cost now threatens to enforce the abandonment of coal-getting capacity already paid for.

Unless there is careful planning, with due regard to the

uncertainties of the future, the same will happen with Western Europe's natural gas. Its contribution to Europe's indigenous fuel supplies could, of course, be a most welcome boon. But if the immediate effect of natural gas flowing into consumption were a further contraction of Western Europe's coal industry, the net effect would be a further weakening of Europe's future position. Natural gas, once discovered, is obviously cheaper to get than coal. But it does not follow that it is really economic to allow gas to replace coal when there is every reason to believe that both gas and coal will be needed later. The publicly-stated size of the Dutch natural gas deposits is equivalent to about twelve months of Western Europe's current fuel consumption. Such gas deposits would have to be discovered every year to meet the annual growth of Europe's fuel requirements.

10

From The place of coal in the future. Address to Combustion Engineering Conferences in Manchester, January 25, and London, February 17, 1966[1]

It has been said that, in view of the recent find, the North Sea might become another Texas.[2] The quantities of the BP contract with the Gas Board, expressed in terms of cubic feet of gas, would appear enormous to the layman but expressed in terms of coal equivalent it is seen as about half a million tons of coal equivalent a year. In the United Kingdom alone the additional fuel requirements over the next fifteen to twenty years are estimated at 200m. tons of coal equivalent so that new gas discoveries under the North Sea would have to be very large indeed to affect the situation. I am not saying that 100m. or 1,000m. tons of coal equivalent might not be found one day; but that even such a discovery would do no more than make an urgently necessary contribution to the fuel supply problem of Britain's expanding economy.

The history of the great gas discovery in Holland illustrates the uncertainty of the situation. Drilling commenced soon after the war; gas in commercial quantities was found in 1959. This aroused great enthusiasm, but by 1965 − six years later −

hardly any of this gas had found its way into ordinary consumption. However, the gas is now beginning to flow and the Dutch expect it to flow to the extent of 30m. tons of coal equivalent a year in 1970. Thirty million tons of coal equivalent a year eleven years after the discovery of what is claimed to be the world's second largest gas field! Britain's fuel requirements alone may well increase by 100m. tons of coal equivalent in eleven years![3] It is only when we quantify the matter that we can see it in proper perspective; doing so, we realise that the real problem is not one of the new fuel displacing the old: the real problem is whether new and old together will be enough to meet the rising requirements.

There has been a very intensive nuclear energy programme in the last five or seven years but it has been totally uneconomic, and the latest stations that have come on stream are still producing power at a cost which could be compared with a colliery that made a loss of £5 per ton of coal. The justification for this vast expenditure, in spite of this total lack of viability, was that we did not want to make ourselves unduly dependent on imported oil, nuclear energy being a semi-indigenous fuel although we have to import the uranium. This thinking in 1955 was quite sound but what has happened? These vastly uneconomic stations have been integrated into the British fuel economy in such a way that they did not reduce Britain's dependence on imported fuel supplies but worked to the detriment of the coal industry. Instead of using a colliery that has already been paid for, society has spent a vast amount of money on a totally uneconomic nuclear reactor, thus simply substituting one fuel for another. The outcome has been, not that Britain's dependence on imported fuel has been reduced, but that many additional collieries have had to be shut.

The insurance policy that Western Europe, and particularly Britain, can take out is to approach its one important indigenous fuel supplier, the coal industry, in the spirit of a good householder. I suggest that this insurance policy would be very cheap indeed, the common people would not even notice its cost. But unfortunately, as a society, we are doing the opposite; we are telling the public, and the coal miners, that coal is a dying industry, or at any rate a declining industry.

97

It is easy for economists and planners to say that we do not want 200m. tons of coal a year, that we only want 175m. or less, and that the social effects of pit closures will be mitigated by generous redundancy arrangements. Redundancy arrangements are not a good sales point in recruiting young miners or holding older ones. Recruitment has declined to the lowest level since nationalisation and voluntary wastage has risen to the highest. How are we going to stop this exodus? Are we going to restore confidence in the industry or are we going to slip into a position where we cannot live without the industry and cannot live with it, and have a sick industry on our hands as it has happened before? This is our big problem, and I think instructed public opinion can help immensely by spreading the understanding that, although at the margin there can be argument about 10m. tons more or less, the real problem is how to keep the bulk of the industry sound.

From a position of real inferiority, the British coal industry has worked its way, in the last nineteen years, to a position of technical superiority — internationally speaking. It would be a pity if that work should now be lost again — not because the engineers had failed us but because our national fuel policies had been incapable of coping with a temporary phase in the world fuel situation, a phase of knock-down oil prices, which the oil companies are the last people to be happy about, a phase when there are already straws in the wind indicating that the situation can soon change again as dramatically as it did in 1957.

We should not lose all the opportunities we have in our only indigenous fuel industry because of a very temporary glut in world fuel supplies.

11

From 'Coal and its competitors', *Colliery Guardian Annual Review*, 1968[1]

More than 21 years have elapsed since the British coal industry was nationalised on January 1, 1947. For the first half of this period, there could never be enough coal and output targets were put up; for the second half of the period, there was too much coal and output targets were put down and down.

The change from a sellers' market to a buyers' market, which took place in 1957 (although not clearly recognisable at that time), followed the first Suez crisis and the restrictions which the United States put on the importation of oil with the purpose of protecting its indigenous oil production against cheap oil from the Middle East. These restrictions, among other factors, led to an oil glut in Western Europe, resulting in oil prices against which European coal found it hard, and often impossible, to compete.

The difficulties of the British coal industry during the last ten years or so were caused by oil competition, and not by the 'new fuels' — nuclear energy and natural gas — which so far have had only a very limited impact.

Things would be easier if collieries could be 'mothballed', but they cannot, and even if they could there is no possibility of mothballing the men needed to work them.

In other words, the coal industry is in a peculiarly difficult position when faced with a decline in demand. Decisions to abandon long-term capacity are virtually irreversible and cannot, therefore, be taken unless there is a well-founded conviction that the decline in demand is also irreversible.

The White Paper on Fuel Policy (Cmnd 3438), unfortunately, deals with these matters only in the most cursory fashion. It expresses the opinion that 'the surplus of crude oil supplies seems likely to persist for many years despite the expansion in world demand', without defining the words 'many years' and without dealing with the evidence that seems to point in the opposite direction. As regards oil prices, the White Paper expects 'that pressure to force up crude oil prices will be held in check by the danger of loss of markets', because 'here as elsewhere oil will be up against increasingly strong competition from natural gas and nuclear power' (paragraph 53, page 26).

By 1970, nuclear energy will be available on a scale equal to 14m. tons of coal a year, provided that the 5,000 Mw programme of the Magnox stations has been completed by the end of

1969. At present, it seems likely that there will be some delays, so that the availability of nuclear energy may be 1 or even 2m. tons of coal equivalent less than the figure given in the White Paper. But this is relatively unimportant. The important point is that the Magnox programme as a whole has the effect of permanently reducing coal demand by 14m. tons a year, although it is totally uncompetitive on strict commercial grounds. Once the Magnox stations have been built, it is of course more economic to use them — and to use them for base load at maximum capacity — than to let them stand idle.

The second nuclear programme, to be based on Advanced Gas-cooled Reactors, is to be even larger than the first. By 1975, the contribution from nuclear energy is expected to amount to 33m. tons of coal equivalent,[2] and again, whether the total costs are competitive or not, once the stations have been built, it will be economic to use them at maximum capacity.

As the White Paper puts it, 'the choice between nuclear and conventional generation rests mainly on estimates of the advance of nuclear technology on the one hand and of the future movements in the cost of conventional fuels on the other. No commercial scale AGR is yet completed and operating, and in a young technology the risk of disappointment must exist' (paragraph 33, page 17).

The 'competition' between coal and nuclear energy is therefore not a matter of actuals but of estimates; the 'actuals' are wholly in favour of coal and oil. With a new technology, this is inevitable; but it does again raise the question of speed. Hitherto, Britain has forced the pace of building nuclear power stations more than any other country in the world — with the result that at the end of 1967 about 40 per cent of the world's nuclear power station capacity was operating in the United Kingdom — and it is now generally recognised that we have paid dearly for it, both in terms of avoidable power generation costs and in terms of costs arising elsewhere in the economy. The fear that the same may be happening again cannot be dismissed out of hand.

In view of these serious uncertainties, the White Paper insists that 'apart from straight comparisons of costs of generation, other considerations must be taken into account by the industry (*sic*!) in assessing the merits of adding one type of station rather

100

than another to the system.' What are these other considerations? One is that 'Nuclear power stations cause no air pollution'. It is true that they do not emit smoke and grit; but it is also true that the process of obtaining heat by nuclear fission causes environmental pollution of a highly ominous kind. This aspect of the matter is hardly ever mentioned in Britain, but it is a lively subject of discussion in other countries.

12

From Coal — the challenging future. Address to Combustion Engineering Association, London, December 4, 1969[1]

Another element which is entering into the nuclear energy field is popular opinion, and in the USA there is now something like a popular revolt. There was a public confrontation recently between what might be called the ordinary people on the one hand and the American Nuclear Energy Authority on the other, and this did not go well for the latter. In America, the Nuclear Energy Authority is responsible both for setting the safety standards and for promoting nuclear energy, and it is becoming increasingly apparent that it is impossible to discharge this dual function to everybody's satisfaction. It is now expected that this dual function will be abolished and that there will be a considerable tightening of standards, particularly on radio-active effluents. This popular revolt is not merely carried on by uninstructed people but by some of the most illustrious professors of biology and other sciences in the USA.

Two years ago the government, undeterred by the Ridley Committee and many other committees, produced a White Paper on Fuel Policy.[2] This has never been debated but it has never been withdrawn, and if one has been in this business for a long time one is not inclined to take its detailed predictions too seriously. However, it has a big influence on public opinion, particularly the public opinion in the coal industry, because that is the industry most immediately concerned.

We are moving towards the first date on which the White

101

Paper gives information on how much coal will be used in the United Kingdom, how much oil, how much nuclear energy and how much natural gas. The White Paper is based on the idea that we are moving from a two fuel economy to a four fuel economy. All perfectly true in qualitative terms but not in quantitative terms, and that is what really matters. The nuclear contribution is now a totally open question. It was assumed that by 1970 the Magnox programme would be complete; it is not complete. Assuming that the Magnox programme would be complete it was assessed that the contribution by nuclear and hydro-power would be 16m. tons of coal equivalent; but now there is a very big question mark against this. It was assessed that natural gas would provide 17m. tons of coal equivalent and again we do not know. In those 17m. tons there is something of the order of 2m. tons from imported Algerian gas and there is no reason to suppose that this will diminish, but it implies an input of 15m. tons from North Sea gas into the United Kingdom market.[3]

No doubt you will have given attention to the formation of the Organisation of the Oil Exporting Countries which is probably the biggest single monopolistic organisation in human history. OPEC was set up in 1960 and at the time it was said that the Arabs could never agree among themselves. The fact is that OPEC controls 85 per cent of all the oil that flows into international trade and has been in existence for nine years. Its effectiveness has been marred by the basic fact that the camp is split among the republicans and the royalists, but the latter have been reduced by the recent happenings in Libya and how long it will be before Saudi Arabia joins the republican camp is a matter of conjecture.

The aim and object of OPEC is to reduce the total flow of oil to the rich countries and to get very much higher prices. After all, their oil is their only asset and if that oil is drained out in the course of the next thirty to fifty years they would have a very slim chance of building up an alternative economy in the meantime. Over the last two decades it has been shown how difficult it is to bring industry into these remote countries with no industrial tradition. This is not a thing that can be done in one development decade; it is a matter of generations. They have taken the cue and have said that they must come to some form of pro-rationing scheme for oil to give them enough time to produce that alternative economy which one day will assuredly be necessary.

In the last financial year the coal industry sold 10·4m. tons more than we were able to produce. At this stage the coal shortage is unnoticeable because we entered the year with a stock of 25m. tons; but by the end of the present financial year, that is to say the end of this winter, our stock level will be lower than it has been at any time in the last decade. Suddenly we find that the position is reversed, and we are not so much worried about sales as about supplies. The problem now is to stabilise output because if we do not we might get into serious supply difficulties.

Ten years was the duration of the crisis. When the first turning point came in 1957 we did not understand why it came, but we do now. I believed in 1967 that we had touched bottom and the second turning point had come. So far since 1967 the total demand has been more or less stable, not only in the United Kingdom but in Western Europe as well. Whether or not we come into a period of expanding coal demand will depend on the performance on many fronts, particularly the cost of nuclear energy and the price of oil.

To my mind it is a marvel that it has been possible to hold a situation of relative industrial peace,[4] but there has been a delayed action effect and we are moving into a situation of much tighter industrial relations. It has played havoc with the reserves of the industry, although this is very difficult to grasp statistically. However, substantial reserves remain and the technology remains. The technology is so new and so potent that our main job is still to learn how to make the best use of it.

13

From *Small is Beautiful,* Blond & Briggs, 1973, pp. 121–3[1]

These warnings, and many others uttered throughout the 1960s,[2] did not merely remain unheeded but were treated with derision and contempt — until the general fuel supplies scare of 1970. Every new discovery of oil, or of natural gas, whether in the Sahara, in the Netherlands, in the North Sea, or in Alaska, was hailed as a major event which 'fundamentally changed all future prospects', as if the type of analysis given . . . had not already assumed that enormous new discoveries would be made every year. The main criticism that can today be made of the exploratory calculations of 1961 is that all the figures are slightly understated. Events have moved faster than I expected ten or twelve years ago.

Even today, soothsayers are still at work suggesting that there is no problem. During the 1960s, it was the oil companies who were the main dispensers of bland assurances, although the figures they provided totally disproved their case. Now, after nearly half the capacity and much more than half the workable reserves of the Western European coal industries have been destroyed, they have changed their tune. It used to be said that OPEC would never amount to anything, because Arabs could never agree with each other, let alone with non-Arabs; today it is clear that OPEC is the greatest cartel-monopoly the world has ever seen. It used to be said that the oil-exporting countries depended on the oil-importing countries just as much as the latter depended on the former; today it is clear that this is based on nothing but wishful thinking, because the need of the oil consumers is so great and their demand so inelastic that the oil-exporting countries, acting in unison, can in fact raise their revenues by the simple device of curtailing output. There are still people who say that if oil prices rose too much (whatever that may mean) oil would price itself out of the market; but it is perfectly obvious that there is no ready substitute for oil to take its place on a quantitatively significant scale, so that oil, in fact, cannot price itself out of the market.

The oil-producing countries, meanwhile, are beginning to

realise that money alone cannot build new sources of livelihood for their populations. To build them needs, in addition to money, immense efforts and a great deal of time. Oil is a 'wasting asset', and the faster it is allowed to waste, the shorter is the time available for the development of a new basis of economic existence. The conclusions are obvious: it is in the real longer-term interest of both the oil-exporting and the oil-importing countries that the 'life-span' of oil should be prolonged as much as possible. The former need time to develop alternative sources of livelihood and the latter need time to adjust their oil-dependent economies to a situation — which is absolutely certain to arise within the lifetime of most people living today — when oil will be scarce and very dear. The greatest danger to both is a continuation of rapid growth in oil production and consumption throughout the world. Catastrophic developments on the oil front could be avoided only if the basic harmony of long-term interests of both groups of countries came to be fully realised and concerted action were taken to stabilise and gradually reduce the annual flow of oil into consumption.

As far as the oil-importing countries are concerned, the problem is obviously most serious for Western Europe and Japan. These two areas are in danger of becoming the 'residuary legatees' for oil imports. No elaborate computer studies are required to establish this stark fact. Until quite recently, Western Europe lived in the comfortable illusion that 'we are entering the age of limitless, cheap energy' and famous scientists, among others, gave it as their considered opinion that in future 'energy will be a drug on the market'. The British White Paper on Fuel Policy, issued in November 1967,[3] proclaimed that:

The discovery of natural gas in the North Sea is a major event in the evolution of Britain's energy supplies. It follows closely upon the coming of age of nuclear power as a potential major source of energy. Together, these two developments will lead to fundamental changes in the pattern of energy demand and supply in the coming years.

Five years later, all that needs to be said is that Britain is more dependent on imported oil than ever before. A report presented

105

to the Secretary of State for the Environment in February 1972,[4] introduces its chapter on energy with the words:

> There is deep-seated unease revealed by the evidence sent to us about the future energy resources, both for this country and for the world as a whole. Assessments vary about the length of time that will elapse before fossil fuels are exhausted, but it is increasingly recognised that their life is limited and satisfactory alternatives must be found. The huge incipient needs of developing countries, the increases in population, the rate at which some sources of energy are being used up without much apparent thought of the consequences, the belief that future resources will be available only at ever-increasing economic cost and the hazards which nuclear power may bring in its train are all factors which contribute to the growing concern.

It is a pity that the 'growing concern' did not show itself in the 1960s, during which nearly half the British coal industry was abandoned as 'uneconomic' — and, once abandoned, it is virtually lost for ever — and it is astonishing that, despite 'growing concern', there is continuing pressure from highly influential quarters to go on with pit closures for 'economic' reasons.

14

From Facing the change in the energy situation. Conference organised by the Royal Institute of British Architects, University of Durham, July 18, 1974[1]

On October 6, 1973, the outbreak of yet another Arab-Israeli war provided the historical opportunity for the Arabian oil exporting countries to establish a radically new situation as regards crude oil prices. Within a few months, these prices quadrupled, and their fundamental aim, which is to conserve their proven oil reserves to last longer than a mere twenty or thirty years, can now be attained *via* the so-called price mechanism. We hear nothing more about 'using oil as a political weapon'; what we shall be hearing more and more insistently from now on is this: 'Of course, you can buy all the oil you want, but how are you going to pay for it?' And the truth of the matter is that we cannot pay for the amount of imported oil which our economies have got used to.

Whether the crunch will come within three months or three years is hard to predict. But I suggest, nothing of any real importance hinges on precision in these matters, certainly not for builders, architects, town planners and such like, whose 'time horizon' is counted in decades, if not centuries. That the days of the cheap-fuel economy are numbered — in fact, that they are over — there cannot be any reasonable doubt. The sooner we realise and accept this, the better will be our chance of adjusting to the new situation without having to go through a period of unimaginable troubles.

The effects of 100 years of cheapness-and-plenty as regards fossil fuels have been extraordinarily far-reaching. The effects of the coming period of dearness-and-scarcity as regards fossil fuels — not necessarily in all parts of the world, but inescapably in Western Europe, Japan, and many of the so-called developing countries — will be equally far-reaching.

15

From Living with the energy shortage. Address to Combustion Engineering Association Conference, Eastbourne, November 27, 1974

I have just come back from the United States where they told me a story of two cows that were grazing peacefully alongside one of the big American highways. A fleet of milk lorries, adorned with slogans, like 'homogenised', 'pasteurised', etc., went past and one cow said to the other: 'Makes you feel kind of inadequate when you see this'. The producer, the real, primary producer, has always been made to feel inadequate. In fact, his rewards have always been kept rather low. Now the people who provide the material basis for everything we do have discovered their power.

Internationally and internally the basic producers have dis-

107

covered their power, and a very clear case is provided by the coalminers of Britain. In the 1960s the coalminers were told that they were not wanted, and it was only due to the kindheartedness of society that all the pits were not closed at once. Under pressure from government, the mass media and public opinion, we were forced to abandon about half the industry and were told that oil would for ever remain cheap and plentiful. No reasoned case against this extraordinary thesis ever made an impression and so, as you know, we had to run down the coal industry even to the point of telling able-bodied miners aged 55 that we would actually pay them to come out of the pits. They were not wanted.

Then, of course, the situation changed in 1970 and the miners were suddenly told that they *were* wanted. Having been very quiet and modest throughout the 1960s, it was not surprising that they now wanted more money. They were told that they would not have as much as they asked for because of inflation; so they threatened to strike. They were told not to be foolish as they would strike themselves out of their jobs; but they were adamant, and three weeks later the lights went out. The result was that the Wilberforce Committee granted them, not only what they originally asked for, but anything that had occurred to them in the meantime.

Spurred by happenings such as this, many other people discovered their power. More and more groups have discovered that they hold society in the hollow of their hand, and the same applies internationally. Do not for a moment assume that it applies only to oil. We all know about OPEC, but although there is no OFEC, no Organisation of the Food Exporting Countries, feedstuff and food prices have also risen, many by factors of three or four. There is no organisation of the phosphate exporting countries but the main supplier, Morocco, realising that theirs is a non-renewable asset, that it is being taken away at a tremendous rate, they ask what is to become of them when it has gone. It is their main livelihood, and so they put up the price of phosphates, essential for modern agriculture, in much the same way as OPEC put up the price of oil.

This is the underlying cause of the great inflationary problems we have, not only in this country but throughout the world. People talk about oil sheikhs, but why not about phosphate sheikhs, or copper sheikhs, or food sheikhs? Since when do we call the Canadians sheikhs? They have also steeply raised the price of the oil they export.

An interesting piece of news, two days ago, was that the Canadians had told the Americans that they were going to phase out their oil exports to the United States. Needless to say this decision on the part of Canada was received with dismay in Washington but I have seen no sign that it was received with dismay in London. Whitehall seems to go on thinking that there is plenty of energy about, for instance, in Athabasca sands and Colorado shales. Of one thing, however, we can be certain: the United Kingdom will not see one ounce of oil from Athabasca sands or Colorado shales in the next twenty years or so. So the sheikhs of Alberta and of Colorado will behave in very much the same way as all the other raw material producers, particularly those who produce non-renewable raw materials.

I hear it said that the energy problem is not a matter of shortage but only of price, and that prices will come down again; that OPEC is a cartel; that cartels never last; and that the Arabs will soon start worrying about maintaining their production. This is just the same as saying that the miners will strike themselves out of a job, and the lights will assuredly go out if we go on believing it. We say that the Arabs must re-cycle their funds — a new expression: but why not call a spade a spade? What we are saying is that we want to have the oil 'on tick', that they should let us have the oil and not demand payment in real goods. This is what re-cycling Arab funds means.

When talking about their fuel policy a few years ago, the Americans announced that in 1985 they were planning to buy as much oil from the Middle East and North Africa as those regions had ever previously produced. Some of us asked what was going to become of us — and Japan — and it was suggested that Saudi Arabia and Iran would kindly double their oil outputs. Sheikh Yamani said they were gratified the Americans wanted their oil but asked how they were going to pay for it. The answer was that they could not actually pay for it but the Arabs could invest their money in America. This led to visions of the Arabs being the main shareholders of General Motors, IBM, Columbia Broadcasting System, etc. When it was suggested to King Feisal that Saudi Arabia would become the owners of America, the King pointed out how easy it was to expropriate the accursed foreign capitalist; with one stroke of the pen they could be expropriated; their oil would be gone, and what could they do? Send a gunboat to Washington?

For two years Dr Khene, the General Secretary of OPEC, has been asking the oil importing countries to mitigate their requirements, because the oil exporting countries could rely only on their proved reserves, and, as far as these were concerned, it was a sell-out. The rich countries may well dream about nuclear energy, North Sea, Alaska, Athabasca sands and Colorado shales, but in the meantime they were taking the Arab oil, which was their only livelihood; and at the rate at which it was going there would be only twenty or thirty years before it was gone, far too short a period to develop an alternative livelihood; 250m. people were involved; when the oil was gone what was to become of them? Back to sand and camels?

Then of course, on October 6, 1973, the fourth Arab-Israeli war started and oil was used as a political weapon. The result was a new level of oil prices, and problems of balance of payments, which are becoming more and more an unmanageable situation for the world as a whole.

The price of oil has increased sixfold since 1970, fourfold since last year, and the total payments to be made for it, if we want to maintain the oil stream as it has been, are far beyond any payment that can be discharged in real goods. It is, of course, very risky to make predictions but I should imagine that twice the 1973 flow of money to the oil exporting countries can, in some form or other, including investment, be absorbed by them. But they have four times the amount and, from this and other indications, I would deduce that we should at least be prepared to live, for the foreseeable future, with only half the imported oil flow that we have been used to. For Western Europe this would mean a cut in total fuel supplies by 30 per cent. This would still leave us with far more fuel than we had in the 1930s, but in the current situation it would be a most onerous and difficult change.

That is why I say the party is over. It has lasted 100 years — a very short episode in the history of man. Short and fateful. I am not saying that everything will stop when the party is over. On the contrary I would say that life will unquestionably go on but

110

it will be a different kind of life. We shall have some fuel, just as we had fuel before the party started. My own hope is that the pressures will not come too fast. I do not know how long the period of grace may be until we have to live with something like 30 per cent less fuel than we have been used to recently. I do not expect that all of you will accept this train of argument, but that is what I am basing myself on.

The idea that the future means the ever increasing elimination of the human factor from the productive process is crystallised in theories with which we have been bombarded for the last twenty years, namely, that the passport for the future is education for leisure! To tell the young that what they have to work hard at now is how to live without work.

Some calculations can be made to show that each of us in this so-called affluent society has 30 energy slaves working for him. Living with higher energy costs means living with fewer energy slaves. The energy slaves are going to go the same way as the domestic servants have gone before them. The people who were able to afford domestic servants did not rejoice in their departure. They found it necessary to adjust themselves, to reconsider their life-style, to redesign their houses or move into other houses. I would ask whether we are not collectively in a similar situation when the energy slaves are becoming very much more expensive than they were when we built all these structures, such as this hotel.

I should emphasise that I am not talking in terms of likes or dislikes, or personal preferences. I am trying to grope into the future to take seriously what has happened, what this immense and hitherto unheard of phenomenon of a fourfold increase in oil prices, and many other prices as well, really means. Just as domestic servants became too expensive, and then ceased to become available, our energy slaves have become very much more expensive and we shall have to learn to live with fewer of them. The great hope I have is that the engineers, and other inventive people, will interest themselves, at least to some extent, in the exploration of possibilities to achieve a different system, to try to find a solution to our problems that are relatively small or much more to the human scale, that are much simpler, and much less capital intensive.

111

5
The Principles of Public Ownership

Introduction

Public ownership of the coal industry became an objective for the unions at the annual Trades Union Congress in 1893: it took more than 50 years of political activity to achieve, despite being one of the foremost aims of the Labour Party since its formation. Sidney and Beatrice Webb were among the earliest of many intellectual advocates. Ownership of the coal deposits remained with the land proprietors, who charged royalties to the mine operators, until the Coal Act of 1938 transferred the reserves to the state on payment of compensation to the owners.

The post-war Labour Government passed the Coal Industry Nationalisation Act of 1946 shortly after it came into office and the whole industry and its assets vested in the National Coal Board on January 1, 1947. Nationalisation of railways, electricity and gas quickly followed.

The Labour Party were intent on extending public ownership to the steel industry and in 1949 passed the Iron and Steel Act, naming February 1951 as the vesting date. Having won a general election in October of that year, the Conservatives set about returning the industry to private ownership in 1953 and almost completed the process, only to see it again transferred to the state by the Labour Government in 1967. It was part of the Conservative case during this long and bitter struggle (in which they were supported by most newspapers), that the industries already nationalised were inefficient and that it would be wrong to add others.

At nationalisation there were about 1,000 collieries, 50 coking plants, 76 brickworks and many other ancillary activities:

112

the Board became the employers of more than three-quarters-of-a-million people and the owners of about one million acres of land and 140,000 houses. The total value of the industry's output was £370m.

This vast inheritance was in poor condition: the industry had a long history of contracting markets, falling output and bitter internal relationships. It had been starved of capital during the war and through the 1930s, when its continuation in private ownership was uncertain.

Schumacher was concerned to see public ownership of the coal industry succeed. In speech after speech he defended the Coal Board against its critics. He thought deeply about the purposes and responsibilities of a nationalised industry; there were technical and economic arguments for public ownership, but social and humanitarian reasons too. Conventional financial tests could not be the end of any assessment of a nationalised industry's results. Conservation policies, particularly in the case of the coal industry, made nationalisation a necessity since private ownership could not give the necessary freedom of choice and decision.

Schumacher's most important writing on the subject of public ownership was unfinished when he died. In 1962 he had written an essay of about 35,000 words, 'Die Sozialisierung in Grossbritannien', for a collection of contributions by different authors, published in a volume edited by Wilhelm Weber and entitled *Gemeinwirtschaft in Westeuropa*.[1] The original essay, which was never published in English, began with an historical account of the development of nationalisation as a political ideal and the motives for it. There followed a long, factual account of how nationalisation had been implemented in the United Kingdom, including detailed descriptions of the form of organisation adopted for each industry, and a discussion of price policy and Parliamentary control.

Much of this material was being radically reorganised by him when he died, with the addition of a good deal more of a philosophical and analytical kind, for publication as a major work in English. There were to have been chapters on 'The Ideology of Private Enterprise' and 'The Concept of Ownership'. He intended too, to elaborate on the need for a 'theory of economic goods' that would recognise the basic differences,

113

first between primary goods (with sub-categories of non-renewable resources, like fossil fuels, and renewable, like agriculture) and secondary goods (divided into manufactures and services).

If he had been able to finish this work, we should have had a book worthy to stand alongside those of Tawney, whom Schumacher particularly revered, the Webbs and G.D.H. Cole. Their writing could be only theoretical whereas Schumacher had twenty years' experience in a nationalised industry whose policy, management and form of organisation he had influenced. A book of the kind he was planning would have been uniquely valuable — the work of a participant who agreed with the philosophy, motives and historical inspiration for public ownership.

1

From 'Die Sozialisierung in Grossbritannien', one of a collection of essays by various authors, *Gemeinwirtschaft in Westeuropa*,[1] later extensively revised

British socialism began, and is only to be understood, as a movement of protest and rejection. It sets its face against the exploitation and pauperisation of the masses (an economic critique), against the degradation of the individual and the deprivation of his rights (a political critique), against the uprooting and debasement of people and their standards (a social and cultural critique), and against the entire system of values of capitalist society (a predominantly religious and ethical critique). The purely technical economic motive, namely the desire to increase efficiency through nationalisation, which at the present time stands so much in the foreground, was originally and until quite recently of negligible significance.

With increasing political success to its credit, the movement of protest and rejection acquired, naturally enough, a more positive ideology whose aims may be summarised as follows: first of all there is the all-embracing, and for that reason scarcely definable, aim of a 'reconstruction of society' in accordance with the ideals of liberty, equality and fraternity, thus to discourage people's egotistical impulses and afford the greatest possible scope to their co-operative and unselfish inclinations. Secondly, there is the narrower, but on that account also more concrete, aim of giving protection and justice to all those who under the existing system are regarded and treated as a means to an end, as a mere 'factor of production', including all the weak and the defenceless who possess nothing but their labour power, and who may often even have lost that. The third aim is closely bound up with the second, although it is at the same time more positive in character: it is not sufficient just to give protection and justice; there must be an actual re-shaping of patterns of life and work, in other words a democratisation of economic and industrial management, so that each and every individual concerned in a production process can have a responsible share in determining the conditions of his productive activity. Two further aims, the fourth and fifth, are more purely technical in character: these may be expressed by the terms 'planning' and 'rationalisation' (that is, greater efficiency). Socialisation, and especially nationalisation, is intended to smooth the way for both these ends, and thereby to contribute to the overall aim of raising living standards.

The comprehensiveness of these aims largely accounts for the strong, indeed passionate, interest of British socialists in all questions concerned with education. Even the contemporary preoccupation with purely economic matters has failed to weaken this interest. The longing for genuine participation in the best cultural life of the nation is perhaps an even more powerful element in the British socialist tradition than the desire for greater material satisfaction. It is a disturbing and entirely recent paradox that with many people this cultural longing appears to have grown steadily weaker as their desire for a higher standard of living has found increasing fulfilment. That is why advocates of socialising measures insist that the economic system must not be judged exclusively by its productive efficiency: the production of integral human beings is more important than the production of commodities.

The term 'nationalisation' is used in Britain primarily and almost exclusively with reference to the great industries organised in the form of public corporations.

The public corporation has been described by Professor William A. Robson, one of the foremost British experts on the subject, in the following terms:

There is no doubt that the public corporation is the most important constitutional innovation which has been evolved in Great Britain during the past fifty years. It is destined to play as important a part in the field of nationalised industry in the twentieth century as the privately owned corporation played in the realm of capitalist organisation in the nineteenth century.

Towards the end of his great work, *Nationalised Industry and Public Ownership,* Robson declares:*

The public corporation is, in my judgment, by far the best organ so far devised in this or any other country for administering nationalised industries or undertakings. Allowing for some teething troubles which are still not entirely cured, the public corporation which we have evolved is an outstanding contribution to public administration in a new and vitally important sphere. It is far better than the joint-stock company owned and controlled by the State; or than government departments engaged in business activity; or than State administrations set up to manage commercial or industrial undertakings, such as those existing in the Netherlands, Scandinavia, France and other countries.

What then is a public corporation? It is a body created by Act of Parliament and subject to the laws of the land just like any private firm. A famous legal judgment in 1950[2] lays down that:

In the eye of the law, the corporation is its own master and is answerable as fully as any other person or corporation. It is not the Crown and has none of the immunities or privileges of the Crown. Its servants are not civil servants and its property is not Crown property. It is as much bound by Acts of Parliament as any other subject of the King. It is, of course, a public authority, and its purposes, no doubt, are

116

public purposes. But it is not a Government department, nor do its powers fall within the province of Government . . .

The characteristics of the public corporation — a body corporate, independently financed, exempt from the forms of parliamentary financial control applicable to government departments, and its employees recruited and remunerated on terms and conditions that the Corporation itself determines — are intended to give 'business flexibility' to the public corporations. In the conduct of its day-to-day business, the public corporation is expected to behave commercially, as a private company. To quote Herbert Morrison[3] (1933): 'We are seeking a combination of public ownership, public accountability, and business management for public ends.' The problem is to combine business flexibility with an adequate measure of public control and general policy, and Robson himself admits that, whilst the public corporation has come to stay, it has created some problems, e.g., its relations with the goverment and the legislature, and doubts if the right balance has been struck between independence and political control.

Right from its inception, the National Coal Board was subject to criticisms levelled at its organisation (and efficiency). Much of this criticism sprang from political hostility to the principle of nationalisation, and the legal centralisation of the nationalised coal industry gave ill-will, combined with ignorance, a valuable target to attack. Equally, however, one must not forget the gigantic task which faced the National Coal Board in creating a new organisation in a matter of months for managing 1,000 collieries (and allied activities) previously run by 800 companies.

Nationalisation is, in the first instance, a negative act, the abolition of private property demarcations; and hence the creation of new possibilities for rational reorganisation. In this sense, nationalisation itself establishes nothing but only undoes what was established previously.

This negative proceeding must be followed by something more positive: the creation of a new form or structure. This is a problem for which — as so often in human affairs — there is no

117

unambiguous or final solution; for it is a matter of combining unity and diversity, discipline and freedom, national responsibility and local enterprise, centralisation and devolution. Since each of these pairs of opposites can be reconciled only 'existentially', but not logically, there is a constant danger that either member of the pair may suffer undue neglect or be held up as 'wrong'. The Labour Government may perhaps be criticised for having concentrated too exclusively upon unity, discipline, national responsibility and hence centralisation, and for having, in consequence, paid too little attention to the problems of the internal hierarchy and thus to existing diversity and the need for local enterprise. Some of these deficiencies have since been made good; but now the main danger seems to lie in the other direction, particularly as regards the transport system — that is, people are tending to notice only the diversities, and to want to split up organisations which do, after all, belong together and constitute a single system in some sense.

The reconciliation of opposites is thus not a task that can be absolutely and finally fulfilled with the aid of strict logic. But . . . skilful institutional arrangements can greatly improve the chances of some kind of 'existential' solution being found (which, however, is still never final, but must be renewed and adapted from day to day). The splitting up of a nationalised industry into a number of more or less autonomous firms is indicated wherever clear regional or functional divisions are relatively easy to draw, as, for instance, in the case of the distribution (but not the generation) of electric power, or the production and distribution of gas. The necessity for unity and co-ordination may then be satisfied by the establishment of some sort of 'federal' body at the national level. Where, however, such regional or functional divisions are difficult or awkward to draw, and where a splitting up into autonomous firms is possible only with the aid of forced and irrational demarcations — as, for example, in the case of the railways or the coal industry — a measure of genuine devolution, not entailing the destruction of unity, is only to be effected through the idea of the 'quasi-firm', that is, by constituting individual parts of the industry *as if* they were separate firms. And the decisive criterion of such a procedure is that each of the individual parts should draw up its own balance sheet.

118

It may not be superfluous to point out that problems very similar to those which we have been discussing exist in present-day European politics. How is European unity to be attained without doing violence in the process to the continent's diversity? How are the wealth and colour of European life and society, which are at once a cause and an effect of its diversity, to be preserved without loss of essential unity? Here, too, the pendulum is liable to swing very easily from one extreme to the other, although it ought really to be clear that only solutions which do equal justice to both viewpoints can be genuinely fruitful.

2

'Price policy of nationalised industries'. Unpublished paper, March 10, 1955

It is often suggested that nationalised enterprises should be operated on a no-profit basis; that, indeed, it is one of the advantages of nationalisation that the consumer is able to obtain the products of such enterprises 'at cost', that is, free of profit.

The first thing I should like to emphasise is that, while the word profit, with some people, may have a bad sound, this sound attaches only to *private* profit and cannot possibly attach to any profit of public enterprise. That is, because what is often considered objectionable is the *private appropriation of wealth* through profit-making; but in the case of *public* enterprise there is no private appropriation of wealth at all. If, for instance, the National Coal Board in the United Kingdom would sell its coal at £1 per ton more than cost, the general public, as coal consumers, would 'lose' £200m. (on 200m. tons of coal), but the same general public, as taxpayers, would 'gain' exactly those £200m. The 'profit' of £1 per ton of coal, in other words, would be nothing but an indirect tax on coal consumption. It may be arguable whether coal is the right thing to tax or not, but this is an argument about taxes and not — as ordinarily understood — about profits.

Whether or not nationalised enterprises make profits, the public always and inevitably obtains the services of these enter-

119

prises free of profit, at cost — that is, without any private appropriation of wealth. (Such phrases as 'It is objectionable for the State to exploit its monopoly' entirely miss the most important point, namely, that the word 'exploitation' can be used only in connection with the private appropriation of wealth, but never in connection with a public profit that is essentially a tax.)

Now it is necessary to go one step further and realise that to sell the products of nationalised enterprises free of profit, at cost, means in fact to subsidise them. How could that be? Private enterprise insists on charging a 'normal rate of profit' and will, in fact, not stay in business if such a rate is not forthcoming — so much so, that some economists even count the 'normal rate of profit' as part of production costs. If the products of nationalised enterprises are sold at cost, without a normal rate of profit, they are sold relatively too cheaply, exactly as if a product of private enterprise received a subsidy. The so-called entrepreneurial function, for the discharge of which private enterprise claims to be entitled to a normal rate of profit, would be supplied by the government free of charge, and that is just as much a subsidy as if capital, land, or labour were supplied free of charge.

What is more, the government, by nationalising a certain sector, would automatically curtail the country's 'taxable capacity', as well as its 'capacity to save'. For what would have happened if that sector had not been nationalised? Private enterprise would have charged prices designed to yield at least a normal rate of profit. Part of the profit would have gone to the State as profits tax; part of it would have been retained in the firm (capital formation); and part of it would have gone as dividends to the shareholders. Now, with the State taking over and selling free of profit, there would be no profits tax and no capital formation, but the whole of the sum in question would go to the consumer — just as if it were paid out as a subsidy.

In short, if nationalisation automatically means no-profit operation, the growth of nationalisation inevitably means a shrinkage in the country's capacity to pay taxes and to form capital.

After many generations of capitalistic working for profit, it means assuming a heavy psychological burden if one now attempts to run State enterprise without profit. People are used to looking on 'profitable' as indicating efficiency and 'unprofitable' as indicating inefficiency. It is impossible to convince the private enterprise community (which does not particularly wish to be convinced in any case) that the absence of profit in public enterprise means anything but inefficiency. Nationalisation in Britain teaches this lesson only too well. The nationalised enterprises are somehow expected (a) never to charge more than cost of production and (b) to show a substantial profit at the end of the year. To insist on (a) and then to bemoan the fact that (b) does not materialise has emerged as the perfect tactic to discredit public enterprise in Britain.

Equally important are the internal effects. The people inside the organisation are as much in the grip of the 'profit ideology' as the people outside. They get confused. If it is official policy not to make profits, does that mean that 'costs do not matter', that there should be 'output at any price'? They do not know how to judge capital investments and so forth. These are not logical difficulties, but psychological ones. No doubt, in time people will get out of their long-formed habits of thought, but it will take a very long time, and there is no compelling reason for saddling public enterprise with these psychological burdens now.

Finally, it must be realised that the strict operation of the no-profit rule will in actual practice tend to produce small profits during half the years and small losses during the other half. In the years which − by the operation of pure chance factors − end with a small loss, management tends to become panicky and to make all sorts of destructive 'economy cuts' which disturb steady development. And public relations will be difficult at the same time. Again it must be emphasised that these are not theoretical speculations but matters of actual experience in nationalised, no-profit concerns.

The first principle of price policy for nationalised enterprise, therefore, should be to strive to maximise profits, subject to tax exactly like private profits. This principle, of course, requires to be modified in a commonsense manner where the enterprise is in a monopoly position. In that case, the considerations that apply are simply considerations of tax policy. Is the product of the enterprise a good thing to tax or not? On the whole every-

121

thing is a good thing to tax unless there is definite proof of the contrary. (Such proof must establish not simply that it is 'bad' to tax this particular article, but that the same revenue could be obtained more painlessly elsewhere.)

It is often believed that 'basic' goods like steel, fuel, building materials, etc., or services like transportation, have a particular claim to be left untaxed and to be sold at the lowest possible price. This notion is generally based on error. It is true, of course, that if one article is taxed more than another the demand for the former will tend to be more curtailed than the demand for the latter; and it may be in the public interest that some goods should be in greater demand than others. But this is precisely an argument that cannot be applied to 'basic' goods or services. These enter into everything — that is why they are called 'basic'; and everything cannot be in greater demand than everything else. The demand for 'basic' goods is generally *derived* demand, that is to say, nobody wants these basic goods — like steel or cement — for their own sakes, but only for the sake of some further production.

This matter is perhaps more easily grasped by considering subsidies. After all, a subsidy is merely a negative tax, and a tax is a negative subsidy. Why not subsidise 'basic' materials, transportation rates, capital goods, foodstuffs, etc.? Because nothing would be gained by it: very large sums of money would have to be taken from the public by various forms of taxation in order to be returned to just the same public in the form of subsidies. People say that it would mean taking from the rich and giving to the poor; but this is not correct: it always means giving to everybody, whether rich or poor. If it is public policy to take from the rich and give to the poor, this can be done directly; it is always cheaper to give to those in actual need than to give to everybody in case there are some in need among them. But a subsidy on any 'basic' or 'essential' good or service means, in effect, giving to everybody.

Exceptions do, of course, exist. For example, an industry with large overhead costs — like electricity supply — is often perfectly justified in selling, for an initial period, below actual costs of production, that is, at a price that will cover production costs only when an adequate volume of sales has been built up.

Such cases, however, are not difficult to recognise and are unlikely to be frequent.

I conclude, therefore, that, as a general rule and within certain reasonable limits, nationalised enterprises should strive to maximise profits, so as to make the greatest possible contribution to public revenue and capital formation. Deviations from this rule should be treated as exceptions requiring very special justification — such as: large proportion of overheads and need to build up 'load', or strong reasons of public policy such as would suggest subsidisation in the case of private enterprise.

3

Coal and inflation. National Coal Board press release, September 9, 1957

Mr E. F. Schumacher, the Economic Adviser to the National Coal Board, speaking at the Board's Summer School in Oxford today answered critics of the coal industry who claimed that increased coal prices were the main cause of the inflationary spiral. Mr Schumacher said that such critics usually argued that an increase in coal prices produced a consequential increase in other prices, such as freight charges, which increased the costs of the coal industry and made further increases in price inevitable. This argument of mutual stimulation would not stand up to quantitative examination.

A 10 per cent increase in the price of coal, even assuming that coal accounted for 10 per cent of the national income, would cause a general increase in the costs of industry of 1 per cent. To meet this increase in costs a further price increase of 1 per cent would be required. The total price increase required to account for all the consequential increase would never reach $11 \cdot 2$ per cent. 'It is not an endless spiral', he declared. 'It is only the arithmetic which is endless.'

He pointed out that there was a period between 1948 and 1950 during which coal prices were not increased. But during that period the general price level increased by 20 per cent. Whatever caused that increase it could not have been the price of coal.

'Inflation is a matter where everybody is involved', he said. 'Everybody suffers from the actions of everybody else. The weight that everybody has is measured by their contribution to the national economy.'

Coal's contribution was 4 per cent.

'I am not saying that it is not serious, but in terms of inflationary effect it is a small tail of a very big dog', he said. 'If you come to food prices, which have risen 87 per cent while fuel and power have risen 78 per cent, that is a much bigger tail wagging the same dog, for the weight of food in our expenditure is about six times that of coal, power and light.'

Any change in the price of food had an inflationary effect which was six times as great as the same change in the price of fuel, power and light.

Mr Schumacher also refuted the critics who claim that the Board's investment programme of £1,000m. in ten years was excessive.

He pointed out that expenditure of £100m. a year was about 3 per cent of the nation's total gross capital investment.

'Is that excessive for an industry which provides 86 per cent of all the primary fuel of this country?' he enquired. 'Coal will continue to be the lifeblood of the economy, and who can call it extravagant to invest in it at this rate?'

He pointed out that much of the industry's capital expenditure was being found internally and only some £35m. from the Treasury.

'How much is that really? It is about equivalent to this country's weekly expenditure on armaments. When you put this £35m. into the context of the whole economy of this country it is a small amount of money.'

'The destructive criticisms which we hear and read of the industry are very destructive indeed', he declared. 'Because 99 per cent bears no relation to the facts, we could afford to ignore it but for the effect it has on the morale of the 800,000-odd people working in the industry who take it in sub-consciously.'

4

From 'Britain's coal', *National Provincial Bank Review*,
November 1957

Coal is the only source of power — and indeed the only raw
material — of which Britain has abundant reserves. In this
article I propose to describe some of the things that the National
Coal Board have done, and are doing, to create a modern and
efficient coalmining industry, capable of contributing its full
share to the rapidly growing energy requirements of this country.

Looking back over ten years, there is little point in dwelling
on the unsatisfactory condition of the industry in 1947.
Nevertheless, the industry's performance is not independent of
its past; and it must be remembered that the Board took over
responsibility for an industry which had behind it a long history
of depressed markets, declining output and bitter external and
internal controversy. Bearing this in mind, I think it can be said
that the Board can be proud of what has been achieved during
the first ten years of national ownership.

The primary purposes of the National Coal Board as defined
in the Coal Industry Nationalisation Act, 1946, are: to work the
coal, to secure the efficient organisation of the industry, and to
make available supplies of coal of such qualities and sizes, in
such quantities and at such prices, as seem to the Board best
calculated to further the public interest. In more concrete
terms, the Board are responsible for the operation of some 850
collieries, 50 coking plants, 76 brickworks and many other
ancillary undertakings. They own about a million acres of land
and 140,000 houses, and employ nearly 800,000 people (about
4 per cent of the total employed population). The value of the
industry's annual output is over £800m.

An idea of the complexity of the administrative and technical
problems encountered can be conveyed by looking more closely
at some of these figures. For example, whereas manufacturing
industry can normally arrange for its productive units to be of a
fairly standard character, operating under controlled conditions, this is certainly not true of coal mining. Owing to radical
differences in geological structure throughout the coalfields,

the historical development of mines, and so on, no one colliery closely resembles another — however similar they may appear on the surface; and in any one colliery, working conditions underground can, and do, alter considerably from one week to another, and even from day to day. The 'average' colliery is an abstraction. For instance, the 'average' colliery produces about 245,000 tons of coal a year; in fact, about one-third of all pits produce less than 100,000 tons, and at the other end of the scale there are about 50 collieries with yearly outputs of between 700,000 and 1m. tons. Seam thicknesses range from under 20 inches to more than 200 inches. And while, statistically, the 'average' colliery has about 17 miles of underground roadways in use, many have in fact three times this mileage in service.

Of the total 796,000 people employed by the Board, about 50,000 are in the administrative, technical and clerical grades; the rest, 745,000, are employed as shown in Table 5.4.1.

TABLE 5.4.1

At collieries	696,800
At coke ovens and by-product plants	8,461
At brickworks	2,775
Other ancillaries and services	36,588
Total	744,624

In creating an organisation to run this vast, widely dispersed, industrial enterprise, the first problem was to group the collieries into units of a size which could be supervised directly by one individual, without at the same time creating too many tiers or levels of responsibility between the collieries and the National Board. It was therefore decided to have about 50 Areas. Each Area comprises a number of units together producing, on the average, 4m. tons of coal a year and has a turnover (at current prices) of about £16m. The average Area, therefore, is roughly comparable in size to some of the larger organisations in the private sector of industry. The Areas are grouped into Divisions, now eight in number, each corresponding broadly to one of the major coalfields.

126

The National Board, consisting of eight full-time and (at present) three part-time members, are formally responsible for everything that happens in the industry; their main responsibilities, however, can be listed as follows: formulating national policy for the industry; negotiating with unions on pay and conditions of service; approving capital expenditure in excess of £250,000 on any one colliery scheme; supervising coal distribution; and operating a number of common services. The Board is functional in character; that is, each full-time member, except for the Chairman and Deputy Chairman, has a special responsibility and interest in one or more of the main aspects of the Board's activities. There is now a member representing each of the following main functions: Production, Marketing, Finance, Scientific, Industrial Relations and Staff. Members' responsibilities do not extend to day-to-day matters which are the concern of directors-general of the various departments.

It was only to be expected that experience should bring changes in the organisation, and in the relationships between its constituent parts. Nevertheless, nine years after the Board was formed, an independent committee set up to investigate the industry's organisation — The Fleck Committee — was able to report: 'We are satisfied that the main structure of the organisation — Headquarters, Divisions, Areas and collieries — is sound. And so is the principle of "line and staff" on which the organisation is based.' At the same time, the committee suggested a number of ways in which the main organisational structure might be strengthened, and most of its recommendations have since been put into effect.

Given that the organisation is sound and suited to its tasks, what has in fact been achieved since 1947? In general terms, I would argue that the industry is back on its feet for the first time in forty years; that the forces of decline have given way to forces of expansion that are now becoming apparent. To be more precise, I shall try to answer the question: What has the Board to show for the capital it has put into the industry during the past ten years?

Between 1947 and 1956 the Board spent £558m. on capital account. To begin with, this sum represents gross, not net investment. Unlike a factory, a pit is always on the move, and whereas the distinction between replacement and new capacity is a fairly clear one in manufacturing industry, it is almost impossible to make in coal mining. Secondly, the Board as an employer, as an operator of coking and by-product plants, brickworks and so on, invests considerable sums not directly connected with the task of coal-getting. Accordingly, it is useful to analyse the £558m. into its component items (see Table 5.4.2).

TABLE 5.4.2

£37m. was spent on housing for mineworkers (in Areas where local authorities could not provide sufficient housing or where none existed at all);

£120m. was invested in ancillary undertakings (coke ovens, brickworks, workshops, etc.);

£230m. went on depreciation and maintenance (running costs incurred in producing, over the 10 years, some 2,200m. tons of coal); and

£171m. was invested in colliery schemes − principally in major colliery reconstructions and new sinkings.

It is to the last item − the £171m. − that we must look for improved performance in terms of greater output and improved efficiency at collieries. Here we must take the time factor into account. It takes about ten years to sink a new colliery, and eight to carry through a major colliery reconstruction. For this reason, a considerable part of the capital invested is still 'in the pipeline', that is, spent on schemes not yet completed. So far (end of 1956) completed schemes accounted for about one-seventh of the total, or £25m.

This £25m. is embodied in thirty-two completed schemes; the output of these collieries is now 3·1m. tons more than it was in 1952. And there is no reason to believe that the return on the £146m. still in the 'pipeline' of reconstruction at the end of 1956 will be any less rewarding.

The improved performance of the thirty-two colliery schemes

that have been completed can be assessed by comparing their results with those of other groups of collieries, namely collieries where reconstruction is planned or is in progress; collieries which are planned to continue without major change for the time being; and collieries expected to close down during the next fifteen years owing to exhaustion of their resources. Table 5.4.3 shows how changes in the output of these four groups over the past four years can be expressed.

TABLE 5.4.3

Output 1956 compared with output 1952 (1952 = 100)	
Collieries where schemes of reconstruction have been completed	122
Collieries where reconstruction is planned or is in progress	100
'No major change' collieries	98
Short-life collieries	80
All collieries	98

This index demonstrates clearly the substantial rise in output achieved as a result of new investment — and, at the bottom end of the scale, the decline in output that is inevitable as collieries approach the end of their useful lives.

Because large amounts of capital are tied up in schemes of colliery reconstruction and new sinkings, substantial economies can be made by speeding up the rate of doing this work, and a great deal of research and development effort is going into finding new and improved methods of shaft-sinking and tunnelling. The industry's plans for the next ten years involve the driving of no less than 3,000 miles of underground roadways through solid rock. Obviously there are attractive rewards to be gained by securing a quicker return on the capital invested.

During the past ten years, too, many modern engineering techniques have been adapted to the tasks of getting and moving coal, and real headway has been made, on the managerial front, with the introduction throughout the industry of up-to-date techniques of management, such as standard costing, method study and planned maintenance.

Progress with mechanisation at the coalface, which poses many complex engineering problems, has been especially rapid during recent years. The proportion of total output power-loaded at the coalface (as distinct from being loaded by hand) rose from about 2 per cent in 1947 to more than 15 per cent, or 36·4m. tons, in 1956. By the middle of 1957 the proportion had risen to 23 per cent, and nearly 1,000 power-loading machines of many different kinds were in use.

Great efforts are also being made to improve standards of health and safety in the industry. Accident rates show a consistent downward trend; between 1947 and 1956, the number of accidents of the kind that have to be reported to H.M. Inspector of Mines fell by approximately one-third. Most collieries now have a safety officer whose primary duty is accident prevention. An efficient medical service has also been built up, to ensure that new entrants are fit, to help the men in the industry to keep healthy, to provide first-aid treatment and to assist in finding suitable jobs for the disabled. The Board now has about 80 doctors and more than 300 nursing sisters in its employment, and operates over 300 medical centres throughout the coalfields.

In the field of management techniques, two developments may be singled out. One is the introduction of standard costing. In coal mining, its application is complicated by the fact that changing geological conditions entail frequent revisions of standards; nevertheless, standard costing is now in operation at all except the smallest collieries and those with a short life, and at all the Board's coke ovens. Another development of considerable importance to an industry which is rapidly extending its use of mechanical aids of many kinds is planned maintenance of machinery and equipment. It is especially important in a colliery, where machines are often being used in confined spaces underground, many miles from the surface workshops, to ensure that as little time as possible is lost by breakdowns and repairs. Schemes of planned maintenance, providing for regular inspection and maintenance of all plant and equipment, and the regular withdrawal of machines for reconditioning, are now in operation or in progress at more than 400 collieries — 100 more than at the end of 1956.

Impressive as these developments are, in coal mining the human element is of overriding importance. Labour costs account for more than 60 per cent of the cost of every ton of coal mined, and the industry is still beset by labour problems of many kinds. No one would attempt to deny that this is the industry's most vulnerable feature. Frequent disputes and unnecessary stoppages, and irregular attendance among part of the labour-force — I emphasise 'part' because about 80 per cent of the men are excellent attenders — continue to trouble it. But in deploring these weaknesses one must not lose sight of the real progress that has been made. Productivity has shown a marked improvement over the years: in 1947, every 1,000 tons of coal brought to the pithead required 931 manshifts, compared with 812 manshifts in 1956. In terms of general efficiency, that the British coal industry has outstripped its European competitors is clearly reflected in the price of British coal, cheaper than that of any other European producer. Even in relation to strikes it is easy to let criticism submerge the facts. Strikes in the coal industry must be seen not only against the industry's deeply ingrained strike habits of the past, but also against the strike 'climate' in the country as a whole. Whereas for many years past, the coal industry was responsible for about half of all man-days lost in British industry, during the past ten years the situation has changed; the man-days lost through strikes in coal mining have been substantially below one-third of the British total. To look at the strike position from another angle, the number of working days lost through disputes during the ten years since nationalisation is about 37 per cent less than that during the ten years up to 1939.

5

From The Board's economic and commercial problems. Address to National Coal Board Summer School, Oxford, September 1958

The idea of efficiency is meaningful only in relation to one's

131

purposes. If one's purpose is simply to make money, one's efficiency can be gauged directly from the profit and loss account. What is the Board's purpose and duty? According to a famous sentence in the Coal Industry Nationalisation Act of 1946, it is always to act in such a way 'as may seem to them best calculated to further the national interest in all respects'.

No firm in private ownership is under such an obligation. The central purpose of the typical firm in private ownership, on the contrary, is always to act in such a way 'as may seem to them — the directors — best calculated to further the firm's own interest in all respects'.

Thus, for the Board it is the public interest, and for privately-owned industry it is the private interest. How important is the difference?

Some people think that there is no difference between private and public interest. They believe in some sort of 'pre-established harmony', although they admit the need for legislation in cases where the harmony is less than perfect. But this is an extreme view which does not stand close scrutiny. Most reasonable people would agree, first, that over a wide area the private interest and the public interest do indeed coincide; and, second, that this wide area is by no means the whole area of business decisions.

The Board, of course, do not serve any private interest. They are not 'managers' trying to win profits for their shareholders. Thus it could be argued that, in the absence of a private interest, there could be no possible conflict between private and public interest. The National Coal Board, *as a public corporation*, could take it as their central responsibility simply to serve the interests of the corporation, on the slogan: 'What's good for the NCB is good for the country'. A public corporation, after all, 'belongs' to the public and not to any private group, and if it is doing well, the public is thereby doing well. If this were accepted, there would indeed be no important difference between the policies of public enterprise and those of private enterprise; both would equally go for profits, but the effect would be different since, in the case of public enterprise, profits never lead to a private appropriation of wealth.

The Coal Industry Nationalisation Act, however, lays it down specifically that the Board is 'to further the public interest in all respects' not simply the interest of the National Coal Board. We have no reason whatever to assume that the two were thought to be always identical.

In other words, over a wide area of decisions the Board must indeed act 'commercially', in the interest of the NCB, just like a private firm acting in its own interest. But all the time the Board must consider 'the public interest', whatever that may be, and in case of conflict must give precedence to the public interest over the interest of the NCB.

Thus, in addition to the ordinary 'commercial' problems and interests which the NCB has in common with all other industries, there are for the Board 'economic' problems and duties with which privately-owned industry would not normally be concerned.

An obvious example is the Board's importation of American coal to meet excess demand at home at a heavy loss to the Board. The public interest took precedence over the Board's own interest. No private company would consider it its duty to import from abroad, at a heavy loss to itself, simply for the sake of supplying all needs of the home market. What happened, for instance, when coal was short in Western Germany, where the coal industry is privately owned? Did the companies import American coal at their own expense and sell it at a loss? Certainly not. German coal was allocated to consumers on some sort of quota system, and it was left to the consumers themselves, when their requirements exceeded the quota, to buy American coal at very much higher prices. Such an arrangement, of course, would be totally unacceptable in Britain.

It can easily be seen, therefore, that efficiency for the Coal Board does not always mean the same thing as it would mean for private enterprise. Nor can it be measured in the same way. Yet, when we have recognised these — and other — differences, we shall do well to remember the identities. Nationalised industry is every bit as much concerned with the attainment of maximum efficiency as is private industry. Once the public interest has been recognised by the Board's adoption of certain specific policies to that end, then — within the framework thus given — the criterion of efficiency must rule supreme. That is to say, I do not believe that the question of the public interest could, or should, be raised by everybody all the time. I think that, within the framework of the Board's policies, our approach to all matters of day-to-day business management should be just the same as that of enlightened private enter- prise: 'Is this a paying proposition? In all the circumstances, is this the *best* paying proposition?'

The need for efficiency is, if anything, even greater in the coal industry than it is in ordinary private enterprise, because there is no cushion of a profit margin to protect us from outside shock. Any general deterioration in business conditions — such as the country has been experiencing over the last twelve months — threatens private enterprise with reduced profits while it threatens the coal industry with substantial deficits.

Some of the measures that are often advocated outside the industry as suitable for dealing with the problem of coal stocks and failing demand are, in fact, quite uneconomic from any point of view. We are told, for instance, that we should get rid of our stocks by means of 'clearance sales' at 'bargain prices'. But this would neither be in the public interest nor could it be undertaken as if the Board were a private concern. Cheaper coal, you might say, is surely in the public interest? It all depends on what we mean by 'cheaper'. Lower costs of produc- tion are of course in everybody's interest; but prices reduced below production costs are not. Such prices would involve the Board in a loss, and there is no one to pay for the loss except the public either as taxpayers or as coal consumers. Bargain prices, therefore, would not be a bargain for the public as a whole. Nor would they be good business for the Board, because the demand for coal would not be appreciably stimulated by them. Nothing would be gained but much disorder would be created to no purpose.

I think these arguments apply not only to the inland market but also to exports. To sell abroad at prices which are both below inland prices and below costs of production would merely create disorder without securing any constructive pur- pose.

All this means that there is no easy answer to the problem of failing demand, and it underlines the need for efficiency all round. It also underlines the need for *elasticity*.

Here we might just pause to reflect on a further point of

134

difference between public and private enterprise. To put it crudely: private enterprise, in a fluctuating market, needs to worry only over producing too much, never over producing too little; public enterprise is equally embarrassed by too little as by too much. Private enterprise can 'play safe', when in doubt, by keeping the market short of supplies; public enterprise cannot 'play safe': it is expected to hit the centre of the target every time. During the years of coal shortage, it would not have been considered appropriate for the Board to issue advertisements saying: 'There is such a demand for our splendid product that you cannot get all you want. Be patient. We are doing our best.' But such advertisements were proudly issued by private enterprise.

6

From 'Is the ownership debate closed?' *Socialist Commentary*, February 1959, under the name Ernest F. Sutor[1]

Many socialists are wondering whether nationalisation is not really beside the point. There seems to be nothing in it but trouble; it does not appear to be working particularly well — so why bother with it? To the extent that this mood is simply due to the realisation that no transfer of ownership as such produces magnificent results, that everything worthwhile has still to be worked for, devotedly and patiently, without losing sight of the higher, non-economic objectives of socialism, to that extent it is a sign of progress in thought. But we must recognise that there is also a mood of disillusionment, if not defeatism — a feeling that nothing is worth doing and nothing needs to be done.

The idea of public ownership arose as an answer to and protest against the idea of conducting the entire economy on the basis of private greed. Yet that latter idea, as Marx well recognised, had shown an extraordinary power to transform the world.

The bourgeoisie, wherever it has got the upper hand, has put an end to all feudal, patriarchal, idyllic relations and has left no other nexus between man and man than naked self-interest . . . The bourgeoisie, by the rapid improvement of

135

all instruments of production, by the immensely facilitated means of communication, draws all, even the most barbarian, nations into civilisation.

(Communist Manifesto)

What, then, is the essential strength of the idea of private ownership? It is its simplicity. The totality of life can be reduced to one aspect — profits. The businessman, as a private individual, may still be interested in other aspects of life — perhaps even in goodness, truth, and beauty — but as a businessman he must concern himself only with profits. That is the essential idea in all its stark simplicity, and its only merit lies in its simplicity. The power of its appeal stems also precisely from its simplicity. Everything is crystal clear after you have 'reduced' reality to one — one only — of its thousand aspects. You know what to do: whatever produces profit; you know what to avoid: whatever makes a loss. Not only is there an absolute clarity about aims, there is also a perfect measuring rod for success or failure — profits. Let no one befog the issue by asking whether a particular action is conducive to the wealth and well-being of society, whether it leads to moral, aesthetic, or cultural enrichment — simply find out whether it pays; simply investigate whether there is any alternative that pays better. If there is, choose the alternative.

It is no accident that successful businessmen are often astonishingly primitive; they live in a world made primitive by this process of abstraction or 'reduction'. They fit into this simplified version of the world and are satisfied with it. And when the real world occasionally makes its existence known and attempts to force upon their attention a different one of its facets, one not provided for in their philosophy, they tend to become quite helpless. They feel exposed to tremendous and incalculable dangers and freely predict general disaster.

As a result, their judgments on actions dictated by a more comprehensive outlook on the meaning and purpose of life are generally completely worthless. It is a foregone conclusion for them that a different scheme of things, a business, for instance, that is not based on private ownership, cannot possibly work. If it works all the same, then there must be a sinister explanation

– 'exploitation of the consumer', 'hidden subsidies', 'forced labour', 'dumping', 'monopoly', or some dark and terrible accumulation of a debit account which the future will suddenly present.

But this is a digression. The point is that the real strength of the theory of private enterprise lies in this ruthless simplification, which fits so admirably into the mental patterns created by the phenomenal successes of science. The strength of science, too, derives from a 'reduction' of reality to one or the other of its many aspects, including the 'reduction' of all quality to quantity. But just as the powerful concentration of nineteenth-century science on the mechanical aspects of reality had to be abandoned because there was too much of reality that simply did not fit, so the powerful concentration of business life on the aspect of 'profits' has had to be modified and even abandoned because it failed to do justice to the real needs of man. It was the historical achievement of socialists to push this development, with the result that the favourite phrase of the enlightened capitalist today is: 'We are all socialists now'.

That is to say, the capitalist today wishes to deny that the one final aim of all his activities is profit. He says: 'Oh no, we do a lot for our employees which we do not really have to do; we try to preserve the beauty of the countryside; we engage in research that is unlikely to pay off', etc., etc. All these claims are very familiar; sometimes they are justified, sometimes not.

What concerns us here is this: private enterprise 'old style', let us say, goes simply for profits; it thereby achieves a most powerful simplification of objectives and gains a perfect measuring rod of success or failure. Private enterprise 'new style', on the other hand (let us assume), pursues a great variety of objectives; it tries to consider the whole fullness of life and not merely the money-making aspect; it therefore achieves no powerful simplification of objectives and possesses no reliable measuring rod of success or failure. If this is so, private enterprise 'new style', as organised in large joint stock companies, differs from public enterprise only in one respect, namely, that it provides an unearned income to its shareholders.

Clearly the protagonists of capitalism cannot have it both ways. They cannot say 'We are all socialists now' and maintain at the same time that socialism cannot possibly work. If they themselves pursue objectives other than that of profit making, then they cannot very well argue that it becomes impossible to administer the nation's means of production efficiently as soon as considerations other than those of profit-making are allowed to enter. If they can manage without the crude yardstick of money-making, so can nationalised industry.

On the other hand, if all this is rather a sham and private enterprise works for profit and (practically) nothing else; if its pursuit of other objectives is in fact solely dependent on profit-making and constitutes merely its own choice of what to do with some of the profits, then the sooner this is made clear the better. In that case, private enterprise could still claim to possess the powerful virtue of simplicity — as discussed above. Its case against public enterprise would be that the latter is bound to be inefficient precisely because it attempts to pursue several objectives at the same time, and the case of socialists against the former would be the traditional case, which is not primarily an economic one, namely, that it degrades life by its very simplicity, by basing all economic activity solely on the motive of private greed.

The whole crux of economic life — and perhaps of life in general — is that it requires constantly the living reconciliation of opposites which, in strict logic, are unreconcilable. In macro-economics, the management of whole societies, it is necessary always to have both planning and freedom — not by way of a weak and lifeless compromise, but by a free recognition of the legitimacy of, and need for, both. Equally in micro-economics, the management of individual enterprises: on the one hand it is essential that there should be full managerial responsibility and authority; yet it is equally essential that there should be a democratic and free participation of the worker in management decisions. Concentration on one opposite — say, planning — produces Stalinism; while the exclusive concentration on the other produces chaos. The normal answer to either is a swing of the pendulum to the other extreme. Yet a generous and magnanimous intellectual effort — the opposite of nagging, malevolent criticism — can enable a society, at least for short periods, to find a middle way that reconciles the opposites without degrading them.

138

Ownership, whether public or private, is merely an element of framework. It does not by itself settle the kind of objectives to be pursued within the framework. From this point of view it is correct to say that ownership is not the decisive question. But it is also necessary to recognise that private ownership of the means of production is severely limited in its freedom of choice of objectives, because it is compelled to be profit-seeking, and tends to take a narrow and selfish view of things. Public ownership gives complete freedom in the choice of objectives and can therefore be used for any purpose that may be chosen. While private ownership is an instrument that by itself largely determines the ends for which it can be employed, public ownership is an instrument the ends of which are completely undetermined.

There is therefore really no strong case for public ownership if the objectives to be pursued by nationalised industry are to be just as narrow, just as limited as those of capitalist production: profitability and nothing else. Herein lies the real danger to nationalisation in Britain at the present time, not in any imagined inefficiency.

The campaign of the enemies of nationalisation consists of two distinctly separate moves. The first move is an attempt to convince the public at large and the (minority of) people engaged in the nationalised sector that the only thing that matters in the administration of the means of production, distribution, and exchange is profitability; that any departure from this sacred standard − and particularly a departure by nationalised industry − imposes an intolerable burden on everyone and is directly responsible for anything that may go wrong in the economy as a whole. This campaign is remarkably successful. The second move is to suggest that since there is really nothing special at all in the behaviour of nationalised industry, and hence no promise of any progress towards a better society, any further nationalisation would be an obvious case of dogmatic inflexibility, a mere 'grab' organised by frustrated politicians, untaught, unteachable, and incapable of intellectual doubt. This neat little plan has all the more chance of success if it can be supported by a governmental price policy for the products of the nationalised industries which makes it virtually impossible for them to earn a profit.

It must be clearly recognised that this strategy, aided by a systematic smear campaign against the nationalised industries, has not been without effect on socialist thinking. The reason is neither an error in the original socialist inspiration nor any actual failure in the conduct of the nationalised industries — accusations of that kind are quite unsupportable — but a lack of vision on the part of the socialists themselves. They will not recover, and nationalisation will not fulfil its function, unless they recover their vision.

What is at stake is not economics but culture; not the standard of living, but the quality of life. Economics and the standard of living can just as well be looked after by a capitalist system, moderated by a bit of Keynesian planning and redistributive taxation. But culture and, generally, the quality of life, can only be debased by such a system. Socialists should insist on using the nationalised industries not simply to out-capitalise the capitalists — an attempt in which they may or may not succeed — but to evolve a more democratic and dignified system of industrial administration, a more humane employment of machinery, and a more intelligent utilisation of the fruits of human ingenuity and effort. If they can do that, they have the future in their hands. If they cannot, they have nothing to offer worthy of the sweat of free men.

7

From Prosperous Britain — where does the coal industry stand? Talk to National Coal Board Headquarters staff, London, November 21, 1960

The lazy, small-minded, cynical reference to vested interests is today the main element of discussion — and how clever it sounds! And how futile it is in reality! When someone says anything of significance at all — for instance, to bring it near home, that it is necessary to 'conserve' the coal industry because the chances are that much more coal will be needed in the 1980s than is needed now — people say: 'Ah well, he works for the Coal Board', or: 'Of course, he says that only to please the National Union of Mineworkers, or to embarrass X.Y.Z.', or

140

they will discover a political, or sociological, or psychological, or even physiological explanation as to how the fellow came to make such a statement. The one really important question, *whether or not the statement is true* — that question hardly comes up at all. In all the din of special pleading, there is a contempt for ideas, a contempt for truth, an unholy impatience 'to get on with the job', whether the job is worth doing or not, whether it leads to good or to evil.

While there is obviously no theoretical limit to the ability of human beings to mismanage their affairs, no serious mismanagement has in fact occurred in this country for twenty years, and all the anguished cries of 'crisis' have been vain and useless agitations. The great public, as a matter of fact, know this in a subconscious way, which is shown by the failure of all the shouting to move them. They have never had it so good and they couldn't care less, which is quite right although it causes alarm and despondency among professional journalists and politicians. To the extent that the political parties continue to try and appeal to the voter mainly by means of economic arguments, they meet increasing apathy; when the parties try to outbid one another with economic promises, they earn incredulous contempt.

Among the unwise things done by the affluent society of present-day Britain, I should put first the general attitude taken to the great nationalised industries. These industries have been taken out of the arena of private enterprise so that they should be able to act in a manner 'best calculated to serve the public interest in all respects'. If it had been the intention of the legislators that the nationalised industries should act in exactly the same way as private enterprise, that is to say, in a purely 'commercial' manner, leaving it to the government alone to watch over the public interest, they would have said so. But then the whole nationalisation would have been pretty pointless. No, it seems quite clear that the legislators intended to create instruments suitable for following higher policies than purely commercial ones, to serve the public interest *in all respects* — and indeed some of the Nationalisation Acts made special mention of some of these non-commercial duties, relating mainly to the welfare of workers and other employees. The pursuit of such higher aims is of course likely to incur some

141

extra cost; but in an ever more affluent society this should not matter, as there could hardly be a better cause to which to devote society's affluence than the systematic humanisation of the conditions of work.

Strange to say, as this society has become more affluent over the last ten years or so, the idea that nationalised industries should be pioneers and engines of progress in all the fields which the rough and tumble of competitive private enterprise tends to neglect — this idea which, as far as I can see, is the main justification of nationalisation as such, has moved more and more out of sight, and the performance of nationalised industry is judged ever more exclusively on financial criteria. With increasing affluence, one might have thought, there would be increasing interest in getting things done which are valuable although they do not pay for themselves, there would be a heightened determination that human values should no longer be sacrificed to economic considerations and that human hardship should not be imposed for economic gain; but this is not so. And I think that is a great pity.

Mind you — and let there be no misunderstanding on this point — no nationalised industry can pursue such 'higher' policies unless in doing so it is fully supported by the government and by the general public. If such support is not forthcoming, it is forced into a purely 'commercial' attitude, whether it likes it or not. If financial profits and low product prices are the *only* things that count, the only values that are recognised, one single industry, even if nationalised, cannot swim against the stream.

I am glad to say that the record of the National Coal Board . . . is on the whole an excellent one. Take the manner in which the industry has come through the great so-called coal crisis of the last three years. The number of redundancies — and this is an essential matter of intense human significance, as each redundancy means a severe shock to a man and his family — in three years was smaller than those perpetrated by the motor industry in a week. The voluntary Saturday shift has been given up and overtime work has been curtailed, but there has been no short-time working and hence no grievous reduction in the men's earnings. Compare this with what has recently been happening

142

in the motor industry and its satellites: tens of thousands of men on short-time work, and this in an affluent society which promotes hire purchase on a grand scale, so that any sudden and substantial drop in earnings may cause the most acute difficulties.

In a society which is both affluent and dynamic, there are bound to be unpredictable changes in demand which cause a certain amount of loss. The question is: who is to bear the loss? Some luckless individuals, to whom it may spell ruination, or society as a whole, which would hardly notice it? Put in this way, the answer appears to be obvious. In any case, the National Coal Board quite deliberately chose policies to distribute the burdens of the coal crisis and not to heap them on a relatively small number of luckless individuals. Instead of dismissals and short-time working, there was (a) the building up of coal stocks, (b) the curtailment of opencast (which is done by machines rather than men), and (c) the control, that is to say, the restriction of recruitment. Pit closures, which have attracted so much attention, have been a very minor element in the policy as a whole; with a few exceptions, where the coal was of an unsaleable quality, closures were confined to collieries virtually exhausted.

This policy, which put men before money, which subordinated commercial interest to humane considerations, has no parallel anywhere, so far as I can see – taking into consideration its size and scope – and constitutes a really hopeful new feature in the modern industrial scene. But it has cost money, and this has attracted a tremendous fury of criticism. Nobody knows – if we *insist* on taking money as the only reality – what a policy of ruthlessness would have cost through the despondency and bitterness it would have left among its victims. But such potential costs are never counted. In any case, the well-being of individual men and their families, and not the profit and loss account of any one organisation, should be the primary concern of an affluent society. What is the meaning of affluence, of a 'high standard of living', if it does not mean the avoidance of hardship to guiltless individuals?

The Board's policies have also demonstrated that the careful protection of the rights of men is perfectly compatible with change. It does not mean a static, stationary, ossified society.

143

All that is needed is to temper the rate of change so that it fits in with the natural movements of society.

It is a sad reflection on the confusion of thought prevailing in our society that the Board has come in for the most violent criticism on precisely those policies which are its greatest achievements — policies with which it not only lived up to its highest obligations as a nationalised industry but also recognised the fundamental truth — which should be completely obvious in an affluent society — that it is wrong to chase after a trivial and temporary economic advantage at the cost of grave hurt to a considerable number of honest and guiltless fellow-citizens.

8

From Some problems of coal economics. Paper to a seminar on problems of industrial economics, Birmingham University, January 18, 1962[1]

An important reason for the nationalisation of the coal industry was the thought that private profit-seeking and competition are incompatible with the rational and systematic exploitation of a non-renewable resource. And so they are. The pressure for pit closures and against all policies of conservation is due to a reversion of thought to the more primitive forms of private commercialism, abrogating the special responsibilities inherent in the administration of non-renewable assets. It is to be hoped that academic economists will develop their science sufficiently so that the need for an equivalence of 'the level of organisation' and 'the level of responsibility' will be clearly exhibited. A fully-developed 'Theory of economic goods' would show that the requirement of responsibility is lowest in the case of secondary goods (and particularly, short-lived consumer goods within the group) which would suggest that this is a field most suited for small-unit private enterprise. When it comes to primary goods, a higher degree of responsibility is required owing to the need for conservation, suggesting the necessity for higher forms of organisation. When the business is carried on by small, private units (e.g. farming) there is unquestionably a need for strong collective organs like co-operatives, farmers' unions, and marketing organisations. When the business demands a more comprehensive type of organisation, there is a strong case for nationalisation, as in the case of coal. Nationalisation is a 'high' form of organisation as it gives freedom of choice and decision at many points where competitive private enterprise finds itself under the anonymous compulsion of market forces. Hence it permits the constant application of conservation policies, provided only that the backing of government and public opinion can be obtained.

Unfortunately, the financial set-up of the nationalised industries in Britain is totally unsuitable for risk-bearing business. This is particularly unfortunate for the coal industry because, owing to the limitations of geological knowledge, the development of coal-getting capacity is — and always has been — a highly speculative business.

There are two highly anomalous factors in the financial set-up of the coal industry. First, the industry is financed exclusively on fixed-interest-bearing loan capital and possesses no 'buffer' of equity capital. Secondly, the industry, having been amalgamated into *one* unit, can never write off unsuccessful investments except out of profits, and is therefore forced, in the absence of profits, to carry forward all losses and deficiencies at compound interest.

Economists have generally overlooked these facts and have therefore drawn entirely erroneous conclusions about the profitability of the nationalised coal industry. What is the effect of the absence of equity capital? It is simply that most of what in normal business appears as 'profit' — to be retained as a reserve or to be distributed to shareholders — appears as 'cost' in the accounts of the coal industry. An unfavourable year, which would merely mean the reduction or the omission of a dividend in ordinary business, means the declaration of a 'loss' in the accounts of the Coal Board. And this is not merely a matter of outward appearances: such a 'loss' is an automatic addition to the Board's interest-bearing indebtedness and thereby attracts compound interest charges. The disastrous financial effect of such an arrangement may be understood if I mention that the 'loss' made by the Board in 1947 (the first year of nationalised coal) amounted to £23·3m. As a result, the Board's interest-bearing indebtedness has been increased not only by that amount throughout the last fifteen years, but also by the interest payable on it, and by the interest payable on the interest. It may be recalled that at 3 per cent compound interest a given amount increases by 56 per cent in fifteen years, and at 5 per cent by 108 per cent. At the end of 1960, the Board's accounts showed an accumulated deficiency of £77·9m; but the total 'interest' payments made by the Board to the Minister of Power came to a cumulative total of £300m. In other words, if the Board had been financed exclusively on an equity basis, the accounts would have shown a cumulative surplus of revenue over *total costs* of £220m. As it is, 'dividends' (in the form of interest payments) have been distributed to the Government in excess of the cumulative surplus, and the excess, amounting to £77·9m., appears in the Board's books as a 'deficiency'.

9

From 'Coal forges ahead', *Financial Times Annual Review*, July 8, 1963

In 1957, the National Coal Board still had an average of over 700,000 'men on books', not counting supervisory, administrative, technical and scientific personnel, and produced about 207m. tons of coal including 11m. tons from voluntary Saturday working. At present, the number of men on books is less than 530,000, yet the annual rate of output, without Saturday working, is still nearly 200m. tons. In other words, 75 men today achieve the same output which required 100 men five or six years ago. The 'output per manshift' has risen by one-third. This is a good deal better than the 'Neddy'[1] target of 4 per cent cumulative growth per annum.

It has been achieved in spite of a shortening of the length of the shift by one quarter of an hour, from $7\frac{1}{2}$ hours to $7\frac{1}{4}$ hours; a further 'hidden' increase in productivity of 3 to 4 per cent.

This has not been the result of an 'intensification' of the men's work: on the contrary, with the new type of mechanisation called 'power loading', the underground miner's work has become lighter — less brawn and more brain. Two-thirds of the industry's output is now being won by 'power loading'. So much for manpower productivity.

At the same time there are big improvements in the productivity of machines and materials. 'Power loading', of course, required a considerable amount of capital expenditure, and to make the investment 'pay off', every pit in the country, every coalface, had to be given a new look. Does the layout permit the best use being made of the new machines, the new methods of working, the new technology? Each coalface, which may be many miles distant from the pit bottom, requires a great number of 'supporting services', such as underground transport of men, materials and coal. Can the required output be obtained from fewer coalfaces? The daily output per face has risen phenomenally — from 190 tons in 1957 to 350 tons currently — and the number of coalfaces regularly worked has fallen from 4,200 in 1957 to about 2,500 now. This movement of 'concentration' underground has meant a large saving not only in men but also in materials and machinery.

And so much more coal can now be obtained from each coalface (per day or per week), the collieries in total now have considerably more capacity than is needed. There are, in fact, at present more holes in the ground than are required for getting a steady 200m. tons a year. It is possible, therefore, to achieve a further degree of 'concentration' by the closing of uneconomic collieries. Colliery closures, of course, are going on all the time, and have been going on at a steady rate even during the period of coal shortage. When a pit is exhausted, or virtually exhausted, its closure cannot be avoided. When the industry was chronically short of capacity, colliery closures had to be postponed as far as ever possible (which is possible because even in an 'exhausted' pit there is always *some* coal left); but now the industry is embarrassed by excess capacity, and closures have often to take place when the pit's *economic* reserves have gone. Every closure, however, is a major social operation and may impose hardship on a whole community. The Board see it as their task to temper the rate of change so that those of their workpeople who are affected by colliery closures have the best possible chance of finding an alternative job in the industry. If they have to move, every effort is made to provide them not only with a suitable job but with a house as well. This policy helps productivity and pays off in the end.

The closure of uneconomic collieries — some people call it 'negative rationalisation' — is only a minor source of productivity improvement. The main source has to be found, and is being found, in the general run of collieries which are being radically transformed through mechanisation, concentration and reconstruction. As some of the old and 'marginal' collieries are going out, large-scale reconstructions are being completed. The work inaugurated in the 1950s is paying off in the 1960s. The current achievements prove that the work done in the 1950s, so often impatiently criticised for a lack of quick results, has not been done in vain.

These, then, are the four sources of the great productivity advance of the British coal industry during recent years and currently — 'power loading' mechanisation; 'concentration'; negative rationalisation through colliery closures; and the great reconstruction work launched in the 1950s.

On each of these four counts there is still much room for further advance, and the industry is confident that the 'trend-line' of productivity established since 1957 will not lose its impetus for many years to come.

10

From *Small is Beautiful*, Blond & Briggs, 1973, pp. 250–3[1]

Ownership

A number of large industries have been 'nationalised' in Britain. They have demonstrated the obvious truth that the quality of an industry depends on the people who run it and not on absentee owners. Yet the nationalised industries, in spite of their great achievements, are still being pursued by the implacable hatred of certain privileged groups. The incessant propaganda against them tends to mislead even people who do not share the hatred and ought to know better. Private enterprise spokesmen never tire of asking for more 'accountability' of nationalised industries. This may be thought to be somewhat ironic – since the accountability of these enterprises, which work solely in the public interest, is already very highly developed, while that of private industry, which works avowedly for private profit, is practically non-existent.

Ownership is not a single right, but a bundle of rights. 'Nationalisation' is not a matter of simply transferring this bundle of rights from A to B, that is to say, from private persons to 'the State', whatever that may mean: it is a matter of making precise choices as to where the various rights of the bundle are to be placed, all of which, before nationalisation, were deemed to belong to the so-called private owner. Tawney,[2] therefore, says succinctly: 'Nationalisation [is] a problem of constitution-making.' Once the legal device of private property has been removed, there is freedom to arrange everything anew – to amalgamate or to dissolve, to centralise or to decentralise, to concentrate power or to diffuse it, to create large units or small units, a unified system, a federal system, or no system at all. As Tawney put it:

> The objection to public ownership, in so far as it is intelligent, is in reality largely an objection to over-centralisation. But the remedy for over-centralisation is not the maintenance of functionless property in private hands, but the decentralised ownership of public property.[3]

'Nationalisation' extinguishes private proprietary rights but does not, by itself, create any new 'ownership' in the existential – as distinct from the legal – sense of the word. Nor does it, by itself, determine what is to become of the original ownership rights and who is to exercise them. It is therefore in a sense a purely negative measure which annuls previous arrangements and creates the opportunity and necessity to make new ones. These new arrangements, made possible through 'nationalisation', must of course fit the needs of each particular case. A number of principles may, however, be observed in all cases of nationalised enterprises providing public services.

First, it is dangerous to mix business and politics. Such a mixing normally produces inefficient business and corrupt politics. The nationalisation act, therefore, should in every case carefully enumerate and define the rights, if any, which the political side, e.g. the minister or any other organ of government, or Parliament, can exercise over the business side, that is to say, the board of management. This is of particular importance with regard to appointments.

Second, nationalised enterprises providing public services should always aim at a profit — in the sense of eating to live, not living to eat — and should build up reserves. They should never distribute profits to anyone, not even to the government. Excessive profits — and that means the building up of excessive reserves — should be avoided by reducing prices.

Third, nationalised enterprises, none the less, should have a statutory obligation 'to serve the public interest in all respects'. The interpretation of what is the 'public interest' must be left to the enterprise itself, which must be structured accordingly. It is useless to pretend that the nationalised enterprise should be concerned only with profits, as if it worked for private shareholders, while the interpretation of the public interest could be left to government alone. This idea has unfortunately invaded the theory of how to run nationalised industries in Britain, so that these industries are expected to work only for profit and to deviate from this principle only if instructed by government to do so and compensated by government for doing so. This tidy division of functions may commend itself to theoreticians but has no merit in the real world, for it destroys the very ethos of management within the nationalised industries. 'Serving the public interest in all respects' means nothing unless it permeates the everyday behaviour of management, and this cannot and should not be controlled, let alone financially compensated, by government. That there may be occasional conflicts between profit-seeking and serving the public interest cannot be denied. But this simply means that the task of running a nationalised industry makes higher demands than that of running private enterprise. The idea that a better society could be achieved without making higher demands is self-contradictory and chimerical.

Fourth, to enable the 'public interest' to be recognised and to be safe-guarded in nationalised industries, there is need for arrangements by which all legitimate interests can find expression and exercise influence, namely, those of the employees, the local community, the consumers, and also the competitors, particularly if the last-named are themselves nationalised industries. To implement this principle effectively still requires a good deal of experimentation. No perfect 'models' are available anywhere. The problem is always one of safeguarding these interests without unduly impairing management's ability to manage.

Finally, the chief danger to nationalisation is the planner's addiction to over-centralisation. In general, small enterprises are to be preferred to large ones. Instead of creating a large enterprise by nationalisation — as has invariably been the practice hitherto — and then attempting to decentralise power and responsibility to smaller formations, it is normally better to create semi-autonomous small units first and then to centralise certain functions at a higher level, if the need for better co-ordination can be shown to be paramount.

6
Making the Figures Sing
Business Management and the Use of Statistics

Introduction

Schumacher in 1963 added the role of Director of Statistics to his other responsibilities as the National Coal Board's Economic Adviser. Three years later he wrote of 'the somewhat abstract, somewhat arid and never fully satisfactory kind of numerical information called statistics'. Nevertheless he achieved increased recognition for the value of the art and through it contributed richly to Coal Board management policies. He suggested the statisticians should be a little less humble.

In 1966 he pointed out that 43 per cent of the coalfaces then operating produced less than 200 tons a day and together provided barely 10 per cent of total output. At the other end of the scale, outputs of 800 tons or more were being won from 8 per cent of the faces. He recognised the mining engineers' argument that some of the low output units might be needed as standby capacity in case other faces in the small colliery ran into bad geological conditions. But in the case of most of them, he argued, men, capital, machines, and managerial effort were being used wastefully: efficiency should be increased by concentrating on a smaller number of high output faces. This policy of coalface concentration is still being followed and is even more necessary than it was in his day since with the most advanced face machinery and roof supports now in use, it can cost up to £3m. to equip one unit. There were almost 2,000 faces when Schumacher did his survey in 1966 and the output that year was 174m. tons. In the financial year 1979–80, only 649 faces were needed to produce 109m. tons. Daily output a face was averaging 678 tonnes compared to 419 in 1966.

The Coal Board's early form of organisation was in five tiers with the responsibility running from collieries, to sub-Areas, Areas, Divisions and National Headquarters. The National Board kept at the centre the responsibility for national policy, for negotiating with the unions on pay and conditions of service, for major capital investment, for supervising coal distribution, and for determining prices; it also operated a number of common services for the lower formations. Schumacher was a senior Headquarters official but not a Board member. As an adviser he had only a small staff; in fact in a speech in 1963 to the Combustion Engineering Association he said with some pride: 'I started as Economic Adviser to the [Coal] Board with one assistant and a typist and that is my staff today.'

There was, in the early years of nationalisation, not only the political campaign against nationalisation, but also a good deal of criticism, most of it also emanating from the Conservatives, about the form of the Board's organisation. In particular, the new organisation was said to be over-centralised — a criticism that was regarded by the unions as striking at one of the basic purposes of nationalisation: that the profitable and more advanced coalfields should help the older and more intensively-worked areas which, in any case, produced special qualities of coal unobtainable elsewhere.

A number of reviews of the organisation took place, conducted at first by experts from outside the Coal Board, but the most radical changes resulted during the Chairmanship of Lord Robens, one of Nature's instinctive delegaters, from internal working parties of which Schumacher was a member. The five tiers of management were reduced in 1967 to three — collieries, Areas and National Board. There has been no change in the basic organisation structure since then.

The need to release entrepreneurial qualities in the management of a big public corporation was one to which Schumacher returned again and again. He was also much absorbed in finding for managers of publicly-owned industry a substitute for the motive of profit and individual gain. One of the many great questions he asked was: How can you achieve control of this organisation without killing freedom?

153

1

The search for significant measurements in coal mining. Paper to a conference of the Association of Incorporated Statisticians, November 28, 1960, reprinted in *The Incorporated Statistician*, Volume II, No. 2

Statistics, it has been said, is the art of preventing figures from lying. But this is, perhaps, too negative a definition. To me, as an economist, statistics is the art of showing quality or meaning in quantity.

Quantity, as such, has no meaning at all, just as facts, as such, have no meaning. At the same time, meaning alone − a purely qualitative statement − without quantitative determination has no significance for practical purposes. It is the besetting sin of economists − a sin we must always drive ourselves to guard against − to make qualitative statements of impeccable truth yet of indeterminate significance because there is no indication of their weight, their quantity. How often do we hear, for instance, that this or that will be a bad thing because it will raise prices or increase inflationary pressure or burden the balance of payments. All we can reply is: 'Please, without some quantitative indication your statement is meaningless from a practical point of view.' Perhaps the converse is the besetting sin of statisticians: measurement for its own sake, merely because something can be measured, and hence the production of floods of figures, of oceans of low-grade truth in which all meaning is drowned.

When you have to deal with an organism as complex and as large as the British coal industry, both economists and statisticians are especially liable to fall into their respective errors. But if they work closely together their errors may cancel out, and if good fortune smiles on them, they may even achieve something useful.

One of the qualities which we, like all other producers, are anxious to make visible through quantitative indicators and, if possible, to measure is the efficiency of our production units. But what is a production unit? Table 6.1.1 shows a choice. At which point on this scale can we measure efficiency? At which point can we get quantities which will give a meaningful representation of the quality of 'efficiency'? Now this is a very tricky question which can be answered only from experience. I can remember the early days of nationalisation when we were advised from outside the industry — we always are favoured with a great volume of advice from outside, offered with a degree of sureness that varies with the square of the distance from the scene of operations — that the National Board should limit itself to measuring and 'controlling' the performance of the Divisions. Unfortunately, although it is of course easy enough to collect Divisional data, it proved impossible 'to make the figures sing', or, as I put it before, to make quality visible through quantity. There you had the measurements, but you could not judge what they meant. Well then, what about pushing down to Areas? Each Area is a unified business, led by an Area General Manager who is the undisputed boss of the show. To cut a long story short, even Area figures, although a bit more musical than Divisional figures, cannot often be made to sing. They are what they are, but you do not generally know what they could and should be.

TABLE 6.1.1

	No. of units
National Coal Board	1
Divisions	8
Areas	50
Collieries	718
Coal Faces (about)	4,800
Faceworkers (about)	220,000

You may well be surprised at this. Admittedly, the average Area is a fairly large business employing about 12,000 people — but why should it be impossible to 'judge' Area performance on some fairly simple indicators, just as the Stock Exchange 'judges' firms even larger than the largest one of the coal industry's 50 Areas? I cannot go into detailed explanations and must limit myself to simply giving you the word 'geology'. The geological conditions in British mining vary not only from coalfield to coalfield, but also from pit to pit, indeed from coalface to coalface, and — a further dimension of variability — they vary in time: coalfaces suddenly deteriorate; seams thin out and disappear; there are faults, washouts, and so forth.

When we found that we could not make the Area figures sing, we established — with a good deal of reluctance, I may say — the Colliery Profile. This gave us a chance of studying, for the first time, the anatomy of the industry, and there began a determined and sustained attempt to trace correlations, regularities, 'laws', 'strategic factors' — in short, quantitative factors which would disclose quality, which would point to possibilities of useful action or enable the 'higher levels' to judge managerial performance at pit level. On the whole, I am bound to say, the results were negative, which does not mean they were worthless; every pit, it appeared, is of its own kind. The one and only universal regularity is the absence of regularities. The Colliery Profile, on the whole, did not provide pointers to action but only pointers to investigation. This, of course, is very much a second-best, but decidedly better than nothing.

We could not leave matters there but had to press on. Perhaps it was the coalface, rather than the colliery, that ought to be regarded as the unit. Maybe by reaching right down to the coalface we could obtain figures that would sing, i.e. quantities that would show up essential qualities. If we look upon the coalface as the ultimate unit of production, we see all the work done elsewhere underground and most of the work on the surface (all except coal preparation) as supporting services to the work at the face. The cost of these supporting services, taking the industry as a whole, is considerably more than half the total cost of coal production. It was clear that it must be largely determined by the geographical concentration or dispersal of the coalfaces in relation to the pit bottom. Thus we got the concept of 'geographical concentration'. From an economist's point of view, however, this was still too 'physical' and not sufficiently 'economic'. What matters is not the cost of these supporting services as such, but the cost per ton of output which, it was argued, must be largely determined by the number of tons available from each face per unit of time, say, per five-day week: and this put into relation to the geographical concentration. Thus we obtained the concept of 'tonnage concentration' or 'overall concentration', which, it seemed, would get pretty near to the idea of efficiency as such. There were therefore three ratios, each of which could be ascertained without great difficulty:

a the length of coalface per length of coal haulage (geographical concentration);

b the tonnage of coal, per five-day week, from each length of coalface; and

c the tonnage of coal per length of coal haulage (overall concentration), being the product of (a) and (b).

It is always difficult to advance from technical to economic concepts. Geographical concentration, that is, the relationship of length of face to length of coal haulage, is something you can see; the tonnage of coal obtained from each yard of face is also something you can see; but the tonnage of coal per yard of coal haulage − this seems crazy; this looks like a purely artificial play with figures: you cannot see tons of coal per yard of haulage; such a thing does not exist in reality. And yet it is *the* economic reality; it is that which really matters from an economic point of view, since it shows how much there is to pay with and how much there is to be paid for.

It is obvious that an increase in 'overall concentration' − (c) above − i.e. an increase in the tonnage available to pay for each yard of the supporting system, must reduce costs. The next question is how such an increase can best be obtained. To avoid misunderstanding let me say that there is no suggestion here of managing the pits from Hobart House[1] or anything like that. Our task is a purely intellectual one, the advancement of understanding and the careful evolution of signposts to action for the practical man. The most fruitful thing in the world is a valid new concept which by virtue of its truth can serve as a signpost to action.

The concentration figures mentioned so far, although (as it were) reaching down to the coalfaces, are still colliery figures, that is averages which might cover a multitude of sins. In general, it seems, averages are invaluable for descriptive purposes, useful as pointers to investigations, but too inarticulate to serve as pointers to action. We decided therefore to try and have a look at all the 4,800 coalfaces separately: to institute a kind of coalface census, which is now done twice a year. But what is the most significant thing about a coalface? What do we really want to know? The mining engineer is first of all interested in 'output per manshift'; this is a measure of labour productivity and of his technical achievement, but not a very good one as there are easy seams and difficult seams. Another measurement of great interest to him is the 'rate of advance', which gives a significant indication of how well the available faceroom is being utilised. For the economist, however, both these measurements are still too technical and insufficiently 'economic'. What is the use of a very high output per manshift at a face or of an excellent rate of advance, if there are only a few men working at it and if the face is so far away from the pit bottom that it requires extensive — and expensive — supporting services which can never pay for themselves? The economic concept is primarily the tonnage of coal obtained from the face, because everything has to be paid for out of this tonnage. In other words, it is the product of the three dimensions — thickness of seam, length of face, and rate of advance — and not any one of the three, which is the finally significant quantity; still more precisely: it is the product per unit of time; in our case, tons of coal per five-day week. This, I believe, is the most significant concept in the conditions of British coal mining today — most significant from an economic and managerial point of view. I do not mean that the traditional measurements, like output per manshift, could or should be abandoned. No, they are valuable and significant in their own way and must be maintained as subsidiary data; but they should not be the primary objects of attention.

The idea of concentration thus gained a new, additional meaning which lends itself to precise measurement, namely, tonnage obtained from each face per five-day week. It is the industry's aim to rationalise production by 'concentration', which now means: by getting the required output from the smallest possible number of faces. At the same time by means of the coalface census we are able to obtain a frequency distribution of coalfaces by size and to identify the faces which, on account of insufficient size, are *prima facie* uneconomic.

Geological and technical reasons often make it necessary to work low-output faces; but the principle we can establish on the basis of these investigations is this: that the burden of proof is on those who wish to establish and operate a coalface scheduled to produce less, per five-day week, than a certain minimum tonnage.

I said 'scheduled to produce'. This is very important. Mining conditions are so variable that you cannot go by what a coalface actually produces during a particular census week. If you did so, you would always find a substantial number of coalfaces which appeared to be unduly 'small' because of geological difficulties of a temporary kind. It is no use directing attention specifically to those coalfaces. The decisive measurement, therefore, is not the actual but the budgeted output per week, which shows what the coalface is expected to do. If it is expected to do less than a certain minimum tonnage, its right to exist is called into question.

To sum up, we found that it was impossible to obtain a real understanding of the performance of the coal industry unless one was prepared to take the coalface as the unit of output and to study the idea of 'colliery concentration'. Two concepts were evolved, each leading to precise measurements and pointing to facts of fundamental economic significance:

a the concept of 'overall concentration', defined as tons of
   coal, per unit of time, per length of coal haulage; and
b budgeted output, per unit of time, of each coalface.

The two concepts are closely related and complement each other. The required measurements reach down to about 4,800 coalfaces; they can be aggregated, analysed and averaged to any desired extent, without losing significance. They can serve as a pointer to action for colliery management and as a pointer to investigation for the higher levels of control.

2

From Coal prospects. Address to the National Association of Colliery Managers, Newcastle-upon-Tyne, November 22, 1952

We are not alone in this world; there are many other people who are up against the same general problems of management that we are up against. I think any impartial observer will admit the following facts:

First,   very great progress has been made in many industries through the application of detailed work study methods, new methods of cost accounting, new practices of machinery maintenance, new methods of manpower deployment and personnel management.

Second,   there are large parts of the coal industry where very little of all this is as yet practised or being tried out; there are large parts where most of it is quite unknown.

I do not believe that any one of you here would challenge the truth of these two statements. And here, I think, lies our one really bright hope for the future. The coal industry in this country is a very old industry, and there are, geologically speaking, no great new virgin territories to be conquered. But speaking not geologically but in terms of managerial method, a vast virgin territory lies in front of us. We have hardly yet begun to exploit it.

To apply new methods means to run risks. The coal industry, under unified national ownership, is now in a better position to run risks than it ever had been in the past. So again, there is a virgin field of opportunity. But we shall miss that opportunity and fail to fulfil our obligations if no experiment is undertaken unless it commands the support of the compact majority. If any new proposition fails to arouse widespread hostility, you can be sure that there is not much in it.

So we need new propositions which, maybe, will arouse widespread hostility. And we should look with a really open mind at what other industries are doing, to find them. Take, for instance, 'Work Study'. This has much wider applications than those connected only with incentive payment systems. It is based on the simple proposition: 'There is more in everything than meets the eye'. And this 'more' that does not meet the eye of even the most experienced manager can be revealed by having trained people to give their whole time to taking detailed, accurate, and methodical measurements of the actual work flow. It is quite amazing what comes to light when these methods are applied — so amazing that at first sight some thoughtless people might jump to the conclusion that management must have been very poor not to have spotted such possibilities of improvement years ago. But this is of course quite wrong. It is not the manager's job to find such things for himself. He has a thousand other things to do. He cannot spend — and must not spend — his time on such detailed investigations. But I suggest that it is his job to get the chaps who can make the investigations. He is not at fault when great possibilities of improvement are discovered by the application of work study methods, but, I suggest, he is at fault if they remain undiscovered because he has never caused those methods to be applied by fully trained men.

I know the case of a firm, reputed to be fairly efficient, which introduced work study methods and achieved, without capital cost, a 70 per cent expansion of output apart from many other benefits. This firm employed about 150 people, and here is a comment that particularly arrested my attention: 'These results (it is stated) were gained at the expense of an additional office staff of only two people'. Obviously, these two additional people were really worth their salaries. And then the thought occurred to me that if the coal industry did the same, engaging 'only' two people to rationalise the work of 150 people, it would have to engage 10,000 people to rationalise the work of all its 750,000 people. And I felt that there is a big lesson in this for all of us. The mere thought of enlarging the industry's non-industrial staff by 10,000 people makes everybody's flesh creep. Yet this is no more than 'only' two persons in 150. This small firm did it and obtained a 70 per cent increase in output. I should think they did right. It paid the price — really quite a small price — and reaped a big reward. Many other firms have done the same. The coal industry, I suggest, could also do the same.

Productivity and output will be raised only by applying more brainpower, better methods, more methodical methods to the job in hand. But to get all this, you need more staff. And this, I

161

think, is really the crux of the whole matter as far as the coal industry is concerned. On the management side there are far too few people, and most of them are grossly overworked. So no one really has the time to investigate new methods of management in other industries or even to give full attention to what might be going on in other coalfields. New methods are therefore all too often rejected out of hand — with some plausible phrase like, 'That is all very well but it would not do in the coal industry'. And how could it be otherwise? The ratio of management to men is lower in the coal industry than in any other industry, trade or service in the country. Very few people seem to know that this is so; most believe the opposite; some just do not believe it when they are told; and others are proud of it. But I think it is no cause for pride. The greatest achievement of our civilisation is in the field of *method* — scientific, technical, administrative, managerial, and so forth. And what is 'method' based upon except conviction, borne out by experience, that there is always more in things than meets the eye?

3

From Hard facts about coal. Address to National Coal Board Summer School, Oxford, September 1953

The first thing, it seems to me, that must strike anyone who looks at the British coal industry is its size. Many of our problems arise from this factor of size, and so do many of our opportunities.

Let us begin with men. The National Coal Board employs nearly as many men as make up the entire male working population of Ireland or Norway. Not quite as many, but nearly as many. If you include their wives, children and dependents, you are dealing with a population of probably over 3m. people — equal to the population of Ireland or Norway. If you include, furthermore, the population which serves the mining communities of this country, which gains its livelihood on account of the miners spending their wages, you are dealing with a fairly large country within the country.

People sometimes talk about the 'battle for coal'. It is a battle, and the front-line is at the face. The total length of faces worked is equal to the distance from London to Aberdeen. Draw this line across the map and every few yards imagine a man battling for coal. And then imagine this line being pushed across the map at the rate of about 220 yards a year. Over 50 square miles — some 35–40,000 acres — of coal seams are yearly taken out from underneath the ground. There is nothing like coal-getting in the whole of human enterprise.

The coal industry is divided into 50 Areas, each employing on the average 15,000 men. A firm employing 15,000 men is normally called 'big business'. The names of most of these firms are household words. There is a certain constructive kind of pride in 'big business' which insists that only the best practices and methods are compatible with the dignity of the firm; that new ideas or methods developed elsewhere should be known and understood in the firm almost before they have been hatched; that the prominence given by size must be backed by a prominence of progressiveness. This type of 'constructive pride' is more common in the newer than in the older industries, but in some cases reaches its highest and most fruitful form in the very oldest. And I think it is due to the enormous strength — or one might say 'potency' — which is conferred by size *provided that the dangers and difficulties inherent in a large size have been faced and overcome*.

In the end, all real progress comes from experimentation. There is a time for research, exploration and discussion, but sooner or later the moment arrives when the new idea — or whatever it is — must be put to the test of practical application. History shows that some brilliant new ideas had to wait for many decades before they were ever put to that practical test, because there was no one big and powerful enough to run the risk of trying them out. But there are also examples of big and powerful organisations having settled down to a particular routine and being unwilling to use their strength for imaginative experimentation. I trust the National Coal Board will never be counted amongst them.

163

I am sure there are two things that we must guard against at all costs, and these are, first, an attitude of mind that rejects experimentation because of the ever-present need for more coal and lower costs, and, second, the view that nothing is worth doing unless it is to be introduced everywhere at the same time. The second danger is even greater than the first. A new idea which readily commands the adherence of the compact majority is normally hardly worth having. The really productive new ideas, as history teaches, have always looked highly doubtful and sometimes quite mad when the first men came along to put them into practice.

But it would surely be quite useless to embark on difficult experiments without adequate staff to carry them through in the best possible manner. In the end, it is only the deployment of first-rate brainpower that can pour new wine into the old bottles. And any marked progress during the next five years or so can come only from the old bottles — I mean from pits already in existence, not from entirely new sinkings. So I believe we ought to experiment in a perfectly calm and deliberate fashion to find out whether the ordinary existing colliery stands in need of more intensive, more detailed, and more methodical management than it is receiving now.

It might be said that more managerial personnel, more specialist services, and so forth would cost money, and that we cannot afford it. Let us be realistic about this. All salaries of £750 a year or more in this industry cost less than $8\frac{1}{2}d$ a ton of coal produced — $8\frac{1}{2}d$ out of 56/9d total costs.[1] If coal is expensive, it is not because of these $8\frac{1}{2}d$, but because of the other 56s. Who would have cause to complain if the $8\frac{1}{2}d$ were doubled and this had the effect of raising efficiency by 10 per cent? Who would not gladly give one shilling to save six? But whether or not such an intensification of management would indeed increase the efficiency of the whole operation, that cannot be settled by argument but only by experiment.

4

'Efficiency in coal production', *Financial Times,* two-part article, December 31, 1953 and January 1, 1954

Efficiency is not a thing which you either have or have not, but is always a matter of degree. Everybody's efficiency lies somewhere between zero and 100 per cent, and it would be nice to know where one stood on that scale. But this will probably never be possible: there are too many variables and too many incompatibilities. It is more important to know whether one is moving up the scale or down, and how one's efficiency compares with that of the next man doing a similar job.

Efficiency, as a workaday concept, is a quantitative term; we cannot make much use of it unless we can measure it. And the very first thing that must be appreciated when discussing efficiency in coal production is that there is no single figure among all the statistics collected by the National Coal Board that can with any confidence be accepted as a measurement of efficiency.

Labour productivity, of course, can be measured fairly easily. Although there are some difficulties in the definition of terms, output per manshift (OMS) serves as a reliable and useful measure. But labour productivity may rise although efficiency is declining, or it may fall although efficiency is increasing. Equally, some pits may be showing low OMS although efficiently run, while others may be showing high OMS although inefficiently run. And the same applies to any other measure of labour productivity, such as a man's output per hour or per week or per year.

The reason lies in the fact that coal mining is not a process of production but a process of extraction. Every year some 40,000 acres of coal seams are extracted, i.e., nearly half a million acres in twelve years. The extraction takes place in about 900 separate pits (some of which have more than one winding shaft).

That is to say, there are something like a thousand pit bottoms from which the daily attack on the ever-receding coal seams is launched — through which access is gained to those 40,000 new acres of coal seams that are taken out in the course of a single year. Some 14,000 miles of underground roadways are needed to get the men to the coal and the coal to the shaft (and for other purposes) — a length of tunnel which, if laid through the earth, would reach to Australia and back again.

These figures serve to illustrate the essential problem of this extractive industry, which can be stated thus: 'The more you produce the more difficult it becomes to produce more.' History and geology are against you, and they determine what can be done today. In ordinary manufacturing industry it is generally quite legitimate to argue that if 100 units were produced some time ago it should be possible to produce more than a hundred units today. But with coal mining it is almost exactly the other way around; large output in the past establishes a presumption against — and not in favour of — large, easy or cheap output in the present or future.

Ever since Vesting Day — seven years ago — the Coal Board have been searching for a reliable indicator by which to judge the performance of divisions, areas, or pits. None has been found. It is probably safe to say that none exists at present. The general conclusion is that the coal industry today, or any part of it, cannot be judged by 'results' but only by the methods it employs. (Specific methods, of course, can be isolated from extraneous factors like sudden geological changes and can then be evaluated in terms of specific results.)

If the methods are improved, results will be better *than they would otherwise have been*. The words in italics are of the greatest importance. Even with improved methods there is no certainty of improved results, because 'other things' are never equal and, as has been said before, 'in the coal industry you have to run very fast to stand still'. But results will be better than they would otherwise have been.

It is extremely difficult to quantify these statements. Let us take the case of capital investment. An extractive industry is continuously eating up its capacity. Large reconstructions and sinking of new pits have to be undertaken all the time to maintain capacity. There is no doubt at all that during the forty years from 1913 to 1953 investment in the coal industry has been insufficient to offset the natural decline of capacity. Only 16 per cent of current output comes from pits less than forty years old.

166

The Board's capital investment is needed, in the first place, to offset the natural wastage of capacity. In one way or another several million tons of capacity disappear every year, of which only a small proportion are accounted for by the closure of virtually exhausted pits. Only when the rate of investment proceeds at a level adequate to offset this natural wastage can we begin to get the benefits of enlarged capacity. These things cannot be measured with exactitude: but it is doubtful whether this level has yet been reached.

The industry is at present carrying through over 100 major projects of reconstruction and new development, the great majority of which are of a magnitude (allowing for the fall in the value of money) which exceeds anything handled before Vesting Day by any person now in the industry. It is easy to say that the Board ought to be doing much more and to do it quickly. But the final bottleneck lies in the human material available. To change from decline to expansion is a tremendous job. It is possible only by a tremendous input of mental energy. It needs additional men, additional brains. But experienced mining men, capable of doing things they had never before done in their lives, are not to be picked up in the streets.

This was recognised by the Board very early on: the industry needs more high-class personnel, and to get them it must train them itself. But this is bound to take time. A lot has to grow invisibly in the soil before anything green appears on the surface. Before new mining engineers could be trained, young men willing (and able) to become mining engineers had to be found. The Board offers 100 university scholarships a year, and though the scheme has proved valuable, the Board has never yet succeeded in finding suitable candidates for all of them.

When the scholars have completed their academic training they must get practical experience under the Board's scheme for 'Directed Practical Training'. Indeed, it takes many years to make a capable mining engineer. The first groups of young men have now completed their training and are beginning to take a hand in the industry's struggle for progress. But it will take some more years before they will have gained the experience necessary to become fully effective.

Meanwhile the Board are exploring many other things. It is now fairly widely recognised that the coal industry − still

167

mainly a 'manual' industry — can only hope to 'demanualise' itself (if such an atrocious word will be forgiven) by obtaining a larger share than hitherto of the country's best brainpower. This is not just a matter of engineering. Each pit and each group of pits face exceedingly difficult problems of administration, accounting, supplies, stores control, manpower deployment, work study, etc. — not to mention the manifold aspects of human relations with regard to which the industry's past history has not always been happy.

Modern methods of management and operations control, which have proved their worth in many industries, have their application also in the coal industry; but they cannot be simply copied or taken over, because the needs of an extractive industry are in many respects fundamentally different from those of manufacturing industries. This means that talent of a high order is required to effect the introduction of advanced methods.

Few people realise how small are the numbers of highly educated and highly trained people in the coal industry today. The Board can claim credit for having raised those numbers a little bit since Vesting Day, but they cannot and do not claim to have done more than scratch the surface of the problem. It is still true that there are too few people trying to do too much.

They are trying to produce ever-more coal to meet ever-increasing demands and at the same time to work off the arrears of reconstruction and development left by a thirty year crisis. On top of this they are trying to do old jobs in a new and — it is hoped — better way. The National Coal Board are fostering experimentation of all kinds throughout the coalfields; but they are not under any illusion that successful experimentation on the required scale could be done by men who have already more to do than they can efficiently cope with.

This, then, is the most urgent and important task before the National Coal Board: to break the 'talent bottleneck' — to broaden and widen management so that it will become commensurate with its broader and wider tasks: to revolutionise the industry by the input of brainpower rather than musclepower. Much seed has been sown and much is growing in the soil. This cannot be seen by people who will only look at overall results. In the coal industry, results are often beyond the power of man.

168

Efficiency in the coal industry cannot be judged by results but only by the methods it employs — by the improvement of methods it achieves.

5

From *Freedom and control in a nationalised industry*. Verbatim record of a speech to a conference on communications organised by the Industrial Welfare Society, October 28, 1964

Freedom and order: I think that these three words are the shortest possible formulation of the real problem of all social life. By all means you can add 'progress' to it, but whether you are considering education, politics, economics, the law or business administration, this is the problem: how can you have the required ordering of things — co-ordination (call it what you like) and yet retain that creative spark of every person in your organisation which can be expressed only if there is freedom? I beg you to consider these terms, not as merely vague thoughts, but as the crux of the problem, particularly in all large organisations.

If certain types of intellectuals get the upper hand, they can always make a perfectly logical case which is relevant to order: everything is neatly ordered and planned and then, because freedom dies, the creative élan of the organisation is absent. Other intellectuals can give rousing lectures about freedom, but if it is implemented in a manner which causes strain or disorder this quality is completely useless and you get a sterile kind of chaos. With an organisation as large as the Coal Board, for instance, this is the real problem, and the problem on which we have been working for the last seventeen years.

The funny thing about freedom and order is that when you lose one you lose both. When you lose the freedom, the order becomes pointless because it is sterile, but if the thing falls into disorder the freedom is quite pointless since it can never really make anything in this world. Before nationalisation, I think the coal industry had all the freedom in the world but was in such disorder that nothing productive came out of it. Upon

169

nationalisation the danger was that, since Parliament had given the chance of establishing order, that order might become sterile and kill the creative spark of the hundreds of thousands of people working in this great industry.

The Coal Board is a very large organisation. Size is no point of pride with me, but it is necessary, and very difficult, to understand the size of the Coal Board. In terms of employment, it is between five and six times the size of the largest private company in this country, ICI. This is not, as I say, a point of pride, but it underlines the problem: how can you achieve control of this organisation without killing freedom?

The first principle I would advocate when you are faced with such a task is one which has been formulated again and again in Catholic social philosophy. It is called the principle of subsidiarity, or alternatively, the principle of subsidiary function. I will read to you a few descriptive remarks in the Papal Encyclical Quadragesimo Anno. You may be surprised that I should quote from such a source, but this is where I found it. It says: 'This is a fundamental principle of social philosophy, unshaken and unchangeable. It is an injustice, a grave evil and a disturbance of right order for a larger and higher association to arrogate to itself functions which can be performed efficiently by smaller and lower societies. Of its very nature, the true aim of all social activity should be to help members of the social body, but never to destroy or absorb them.'

If you let your imagination play on these words, you can see how applicable they must be to any large-scale organisation. It is an evil if you pull functions up which can be efficiently performed below, but the temptation to do so is there all the time because the logical case always seems to point towards that very thing . . . The principle of subsidiarity is: if they can do the job within their own freedom and responsibility, let them do it; take the risk. This is the only way to safeguard human dignity, but also, in the end, to get the best performance from people.

The principle of subsidiarity involves another kind of thought. Imagine a Christmas tree. Most organisation charts look like Christmas trees. There is the star at the top and everything devolves from that. It is all very symmetrical, beautiful and orderly. That is one way of looking at organisation. I prefer a different picture. Imagine a man at a funfair who holds in his hand a lot of strings at the end of which are a lot of balloons. This is my, and I dare say your, ideal picture of organisation — not everything devolving from the star at the top of the Christmas tree, but some co-ordinator who holds all the balloons together. Each balloon has its own life force and power to rise up, and is held by one man who in my picture is not at the top of the organisation but stands underneath it. To put it into more academic language, this is the idea of running a very large organisation in what can be called quasi-firms. We have done this with all definable activities other than the main job of deep mining.

The Coal Board happens also to be the second largest brick producer in this country and a little while ago we set up a bricks executive, organised as a quasi-firm. The Board is a very large producer of coal products, such as coke and chemicals of various kinds. We have taken that activity out of this vast body too, as a separate quasi-firm which is organised as if it were a separate firm, but there remains the man who holds all the balloons together.

Various other functions can be organised in this way. All the time the question is: can we give this sort of inner cohesion and semi-autonomy which will turn administrators into entrepreneurs? I am not speaking polemically, but I think it is worth thinking about the difference between an administrator and an entrepreneur. The difference, it seems to me, is that the administrator fits into the picture of the Christmas tree. He does not fit so well into the picture of the man holding all the balloons.

What is the difference? There is a different ethos of living. It is not that the one is more legitimate than the other. These are two different functions. Anyone who works in a commercial organisation as an administrator gets certain tasks. If he cannot fulfil what is expected of him and he can show that forces outside his control have made it impossible for him, that is his alibi; he is acquitted. But an entrepreneur cannot do that. If he finds that forces outside his control go against him, he has to dream up something which will offset or compensate for these forces.

We have suffered for a long time because we thought that in a large organisation we could, to a large extent, rely on it that all the more important people would give their best; and my God they do. But they used to give their best mainly as administrators. It is said, 'These are my terms of reference. There is a higher level above me. They will check on what I do. If I have a good alibi from factors outside my control, I am all right.' But we have found more and more that you cannot run a big commercial organisation on this principle. You have to run it on a principle which is cruder and, in a sense, more stupid, namely, this is your obligation and by God you are going to fulfil it. If things go against you, dream something up.

I now move from the first to the second principle, which I call identification. It is not enough to tell these subsidiary quasi-firms, 'You will be assessed. You have to make a profit, or a loss' and give them a profit and loss account. It is a grave mistake to stop simply at the profit and loss account. In addition to the profit and loss account there should be a balance sheet. The balance sheet identifies the physical assets which 'belong' to the manager and for which he is responsible, and the inter-action of the profit and loss account and the balance sheet means that if he has done well his assets grow, if only on paper. If he has made losses his assets are diminished, again if only on paper.

You can say that this is only a paper game, but it is a paper game which simulates the reality. If real losses are made, there is a diminution of substance. People say, 'This idea of a quasi-firm is all right. We have a big transport organisation. Let us draw up our accounts so that we show the profit and loss of the transport service. Maybe we should have internal charges so that the facility is not abused and people know what they are doing when ordering transport.' I do not think this does the trick. The job is done properly if the transport organisation, in accordance with what I call the principle of identification, draws up a balance sheet which means that profits and losses are carried forward.

If this is not done, you have a very strange situation which is familiar to us. Your subsidiary quasi-firm has been given a certain task. It may be that if it fulfils it, or over-fulfils it, or does not fulfil it, but the slate is wiped clean at the end of the financial year, a new task is set and the failure or over-achievement of the previous accounting period disappears in the totality of the main firm's accounts. This is a transgression of what I call the principle of identification. The profits and losses should be carried forward unless there is a deliberate top level decision to wipe them out or take them out of the accounts in the same sense as a firm may be refinanced or may distribute its profits instead of keeping them in its own books.

Having said that, you are left with a problem on which I will not enlarge now − this is the third point of my list − the problem of motivation. On this there is a lot of quite futile talk. There is a lot of political talk into which I do not wish to enter, as if the situation of a person like myself working for a national-ised industry was basically different from that of a person working for a large private industry. Of course, we are all employees and we are all working, not for any kind of private appropriation of profits, but in a professional capacity. Our nationalised industries have shown that in order to get very good value out of people, and to get them to give what they have to give, you do not need what so lightheartedly is called the profit motive − that is, a motive where their remuneration is precisely geared to the performance of the organisation as a whole.

That is why I talk about the principle of vindication. A man must know when he is in the clear or when he is not. That means that there can be only one primary criterion. Again, this may cause amusement among you. You cannot boil down economic life to one common denominator unless it is money. Your quasi-firm must have a certain monetary task. In the coal industry, this is extremely difficult. You cannot simply take the break-even point as the minimum monetary task because condi-tions vary greatly from Area to Area. In certain Areas, a large profit is totally unavoidable, and in others a substantial deficit is equally unavoidable. You have to have a handicap system to establish a kind of zero line, but this complication does not arise normally in industries on the surface which are not up against these quite irrational natural factors and there, probably, the problem is very much simpler.

I come to my last point, which I have called 'the middle axiom'.

This is designed, first, to cause in your minds the puzzlement which Socrates has declared to be the beginning of real thought. By 'middle axiom' I mean this (it is a very intriguing question): you have, let us say, the National Coal Board. There is the Chairman and Board at the top and then it proliferates down to a very large number of colliers. When something happens which may not quite fit — say, productivity is not rising, or the industry is not producing enough coal, or is producing too much — the newspapers say: 'Why do not the Board do something about it?'

It is worth reflecting on what such a Board can do. Let us take it in depth again. What in principle can the Chairman and his fellow members on the Board do? If they have an idea of what the industry should be doing, which is a real idea of how productivity can be raised, they can go in for exhortation and preaching and say, for example: 'We wish method study to be applied in the pits.' They can insist that method study engineers be appointed. But in the industry with which I am connected there is a marvellously developed art of stonewalling against such ideas from headquarters.

One other way of influencing the situation would be government by exhortation. Then everybody says: 'All they can do is talk.' What is the alternative? It is to say: 'We have been preaching this or that doctrine for long enough. People do not take it up spontaneously. We will insist. We will take the matter into our own hands at Hobart House and order them to do certain things.' I do not know whether your different industries are as complicated as the coal industry. In the coal industry, if you order things from headquarters, you are bound to make the most horrible mistakes because in every pit things are different. At the time of the great shortage of coal immediately after the war, there was a marvellous machine called the Sampson stripper. The Government and the Coal Board got together and said, 'We will get this machine produced. This is the solution.' Three hundred Sampson strippers were ordered. By the time that the first batch of them got to the pits it was found they did not suit some of the roof and floor conditions, and they could not be used. Compensation for the revocation of the order cost us a lot of money. At this moment I believe that the total number of Sampson strippers in use is three. This just shows that the other alternative to government by exhortation, namely, a sort of do-it-yourself system, is not 'on' and the criticism, 'They are trying to run the industry from Hobart House', is utterly justified.

Where are we left? If we merely preach, we are accused of governing by exhortation, which does not sound good and is not effective. If we say: 'We will tell them what to do', we are accused, rightly, of over-centralisation. Even after you have achieved the proper set-up in quasi-firms, if you find that you are not moving fast enough and are not progressive enough, you want to infuse a certain dynamism into it which you can do only from the top. If it does not come spontaneously from the bottom, you need something between the two extremes, something which is more definite and biting than mere exhortation. Yet it is not the do-it-yourself-at-headquarters arrangement; it is something between. This is why I call it the middle axiom. That is the real art of management.

Let me give an example. It has been clear for a number of years that the way greatly to improve productivity in coalmining is concentration − not concentration by pit closures, but concentration within the pits. If with a new technology we can get far more coal from a particular face, why continue with so many coal faces, each of which requires a very expensive line of communication from the coal face to the pit bottom? If we could reduce the number of production points to half, one-third or one-quarter, this would be a great rationalisation. This was theoretically clear. But somehow it was not happening, or was not happening fast enough. Headquarters, already under tremendous pressure − I have worked in a number of firms but have never seen any business under such pressure as the Coal Board − has to go faster. How do we do it?

It may be said, 'Let us specify the minimum size of a coal face which is permissible.' You cannot do that. Geology is different everywhere, and for safety reasons and proper strata control you must make a tailor-made job of it. This implies carrying all sorts of small faces which on your statistics look unnecessary but which geologically are necessary. What can you do beyond preaching the gospel of concentration? The attempt to do it yourself and to specify and lay down standards is not 'on'. Also it is not 'on' in my general theory that you need the creative co-operation of your people. On the whole, the greater number of people at lower levels are, in total, cleverer than the few people at headquarters.

The middle axiom turned out to be this: to lay down standards for coal faces to be opened. The average life of a coal face is two years. It should measure up to certain standards, but if it does not we require a statement to be made, in a particular book provided for the purpose by the Area General Manager, that he has had, for certain reasons, to open a coal face although it did not measure up to the standard laid down. He also has to answer a few questions. First, why cannot this face be developed to a proper size? Secondly, if it cannot be developed to a proper size, why work this pocket of coal at all? This is followed by a kind of financial assessment and a few other subsidiary things of more technical interest.

This worked like a charm. The power of the Area General Manager, who is the boss of this quasi-firm, was not interfered with. It was up to him to decide whether or not to open the face. The only obligation was that if he opened a face and it did not measure up to standard he had to do some writing. Either the people in my industry are so shy of writing that they will do anything to avoid it, or, more likely, the very process of raising these decisions to the proper level and to the full light of consciousness did away with the abuse and opened the door for this great development of concentration to which the transformation in the fortunes of the Coal Board is due.

6

From Information for management. Paper to National Association of Colliery Managers Conference, Swansea, July 5, 1967

It is fashionable today to say that good management depends on good information, and it would be foolish to deny the truth of this assertion. But it would be equally foolish to overlook two complementary truths, first, that good information is only one of many preconditions of good management, and not always the decisive one, and second, that it is a very subtle and difficult

176

problem to decide what makes information good or bad.

On the first point, let us agree that of course all action must be based on some real knowledge of the subject matter, and there cannot be knowledge without some sort of information, commonly called experience. When we talk about 'information for management', however, we take this general knowledge for granted and mean something more specific: an 'information system'. Now, the point I wish to make is that the process of good management, the process of managerial decision taking, does not normally start with information in this more systematic sense: it starts with a creative idea. When the idea has been conceived — a mysterious process which cannot be organised or routinised — highly detailed information may indeed have to be called for, so as to enable us to judge whether the idea makes sense or not.

If this is accepted, it immediately follows that the 'information system' must never be allowed to become too systematic; it must be kept extremely flexible and responsive. Ideally, it should fulfil three functions. First, it should be such that it can 'spark off' creative ideas, which means bringing to light things which otherwise would remain invisible — matters calling for remedial action or unused opportunities for constructive action; secondly, it should be capable of speedily verifying (or faulting) such creative ideas or suggestions as are put forward for action; and thirdly, no doubt, there is often a case for a certain amount of monitoring and 'control'. When a new line of action has been inaugurated, one wishes to collect specific evidence on whether it is fulfilling expectations or not. Here again, however, the system must not become too systematic and routinised: when it is clear that the new line of action is right, the collection of evidence can normally stop.

This preliminary analysis has already taken us some distance into our main subject which is to try and explore how to distinguish between good information and bad. Let me say right away that this is a subtle subject on which it is easy to enunciate a few common-sense generalities but very difficult to talk with detailed precision.

Among the common-sense generalities, it may not be a waste of time to mention the following.

First, there is the ever-present activity of Parkinson's Law as

177

it affects statistics. It is easier to give birth to a new return than to kill it when it has outlived its usefulness. Who are to be the assassins to do the killing? High level managers have normally more pressing things to do than to scrutinise the established flow of information with a view to eliminating what has ceased to be necessary; and the humble statistician all too often sees his job only as providing information, not suppressing it. The result is unchecked accumulation at ever-increasing expense. Some firms, becoming aware of this, engage an outside consultant as a 'hired assassin'. This has definite advantages in some respects, but disadvantages in others: the outsider may find it difficult to acquire sufficient detailed knowledge to do the job effectively. A better way, I believe, is to encourage the inside statisticians to become less humble and to give them sufficient status within the organisation so that they themselves can cut out the dead wood.

Secondly, there is the ever-present danger of having so much information that 'one cannot see the wood for the trees'. This is a different point from the one just made. Every one of the trees may be alive and healthy, but there are too many of them. How easy it is to list all the things that 'it might be good to know'! The availability of computers may even tempt people to think that 'too much' does not matter any more, because the computer has plenty of capacity. But this is a double error. Information is costly even when it is computerised; it still has to be raised on the spot, checked, and transmitted to the computer. All this absorbs resources, even if the computer can handle the information in a few seconds. The more important point, however, is this: that an excessive amount of information coming out of the computer obscures the vision of management rather than enlightening it, impeding rather than promoting the flow of creative ideas. So one has to be careful — that is all I wish to say at this stage. The phrase about not being able to see the wood for the trees points to a very real danger.

Thirdly, and still on the level of commonsense, there is the converse danger of information preventing one from seeing the trees for the wood. Totals and averages, ratios, percentages and co-efficients are not real things but mental concepts, and action can be applied only to real things. Here again, it is necessary to point to dangers that can arise from the marvellous computing facilities which are now available. If used without the most careful discrimination, they may deflect attention from the living detail of the real world (the 'trees') while focussing it unduly on to certain abstractions which are no doubt interesting in their own way but do not point the way to specific action.

Can we now go beyond these generalities, useful as they undoubtedly are? To do so, we may find it helpful to distinguish between

information for information
and
information for action

The distinction is necessarily somewhat fluid, but it may prove useful all the same.

Some information is obviously not of a kind that action can immediately be taken on the strength of it. When it is presented to the top management, the resulting minute cannot say much more than: 'The Board took note', perhaps 'with satisfaction'; perhaps 'with concern'. Information that is not for action, but merely for information, may appear as a bit of a luxury, and it is just as well to maintain this suspicion.

A Board of Directors will not wish to spend a great deal of time 'taking note', and it follows from that, that regular 'information' should be sparse and concise, just enough to give the general picture, just covering the general items and most basic factors that I have mentioned. There is always a temptation to allow this kind of information to proliferate, and it is never difficult to make a case for additional detail – not simply Area output, but output of each pit; not simply disposals, but disposals by grades, and so on and so forth. I suggest that this temptation should be resisted. The broadest outline of how the business is going is all that is required for 'taking note', and the quicker one can proceed from taking note to taking action, the better. There is a further reason for adopting this principle. No statistic is meaningful by itself: to become meaningful, it has to be seen in comparison with some other data. Now, normally and instinctively, we make comparisons in time – better or worse than last week, last month or last year. We have all found, however, that such comparisons, while helpful, are often not quite good enough. We do not want to tie ourselves to the past but to attain clearly-conceived objectives for the present and the future. By drawing a graph we can see the present in relation to the past easily enough; but we really want to see the present in relation to our objectives or budgets. Now, we all know how difficult it is to set up realistic and meaningful objectives or budgets. One can spend an awful lot of time constructing elaborate budgets, and then events take their own sweet course and one spends another awful lot of time establishing the differences between budget and actual. Most of this time is wasted.

The course of practical wisdom is to keep this apparatus of figures to the really essential minimum, just enough to show the broad outline of how the business is doing.

Totals, averages, and other 'broad outlines' are no use, because you cannot send a man to a total or to an average, and any action minute must be of the pattern: 'Let Mr. A. go to place B and do C.' 'Information for Action' must therefore be of a kind that such a decision can flow from it. The more precisely A, B, and C can be defined, the better. It is very little use telling a man, let us say, to improve his results. We can assume that that is what he is trying to do anyhow. I suggest, therefore, that, as a rule, general financial information does not qualify as 'Information for Action'. You cannot send a man to a financial figure. You may admonish, exhort, cajole – but all that is still a far cry from action.

In fact, it seems to me, the central problem for management at a high level – that is, at a level not directly in touch with physical reality – is always the problem of action. What can 'the Board' actually do? What can the Area Director and his staff actually do? We talk about 'management by objectives', and it is no doubt one of the functions of higher management to set these objectives. But since they cannot be imposed from above but must be determined by agreement, they are not more than a definition of current expectations, needed for the determination of certain overall policies, about prices, financing and so forth. A 'definition of current expectations' is however something very different from any real programme of action. If I

180

define my expectations I may be a bit more optimistic or pessimistic than the man next door or the man above me or below me, but I am all the same only 'talking' and not yet 'acting'. In other words, a manager may be very successful in meeting his objectives, that is, in fulfilling current expectations, but merely because he never permits anything to be expected that requires creative, imaginative action for its attainment.

The principle of 'management by objectives' is therefore not as straightforward a matter as it may seem. If the objective is simply a definition of current expectations, attainable without any really new, creative action and achievement, it serves excellently as a forecast on which the higher level — in our case, the National Board — can base its general policies. But if the objective is in the nature of an ambitious target — a target which can be hit only by new, and therefore unpredictable, departures from current practices — then its value as a firm basis for national policy determination is impaired. In the former case, 'management by objectives' means safety at the national level purchased by stagnation at the operating levels; in the latter case, with ambitious (and therefore somewhat speculative) objectives having been set for the operating levels, the national level is in danger of basing its general policies on expectations that may not be fulfilled.

It seems to me, therefore, that it is necessary in any large organisation to formulate two sets of objectives for its operating units — one in the nature of a forecast, and one in the nature of a plan. And this distinction is very similar to the distinction between 'information for information' and 'information for action'. The 'forecast-objective' merely shows what is the most likely outcome of operations during the coming period, indulging in neither optimism nor pessimism. It is fully quantitative, but does not attempt to quantify more than a few basic items, such as output, disposals, stocks, and profit or loss. The 'action-objective', on the other hand, is primarily qualitative, showing the specific subjects with regard to which new, creative, constructive action is to be taken. An action plan in this sense does not need to contain many forward estimates: if we know in which direction we are wishing to go into unknown territory, we do not normally help ourselves by writing down elaborate estimates on how fast we might be moving.

Let me give an example relating to machine utilisation. Assume for a moment that we can meaningfully express machine utilisation in the form of 'minutes of machine running time per shift' and that a given Area currently achieves an average of 120 minutes. The Area is now called upon to agree an 'objective' for the coming year. What figure is it going to adopt? Surely, an average of 180 minutes 'ought to be possible'. Is the Area's 'business objective' then to be calculated on the assumption that 180 minutes will, in fact, be achieved? This would produce a highly optimistic forecast of results which, if unattained, would gravely embarrass the National Board. A 'realistic' forecast might be 125 minutes, that is, something pretty close to past performance. Such a figure, therefore, will probably go into the formulation of the 'business objective', and it will be the best figure for forecasting purposes. But if this figure is then taken as a basis for an 'action programme', and if Area management subsequently look upon 125 minutes as a satisfactory achievement, are we not caught in a process of 'planning for stagnation'? Management by objectives means that a manager is entitled to be satisfied and 'at ease' when he is attaining his objective. But no one should be satisfied with 125 minutes of machine running time when 180 minutes, or even 240 minutes, 'ought to be possible' and are, in fact, being achieved in other places.

This means, I suggest, that we need to think very hard about the whole concept of 'management by objectives'. On the one hand, we need objectives which serve as a realistic basis for policy decisions at Headquarters: what I call 'forecast-objectives'. On the other hand, we need, at the operating unit, objectives which take into account the great unrealised potentials and also the gross avoidable deficiencies which undoubtedly exist in the industry: what I call 'action-objectives'. The former are 'for information'; the latter are 'for action'. If we confuse the two, we are likely to get poor information for Headquarters and a lack of dynamic, creative, ambitious action at the operating units.

The two different types of 'objectives' necessitate two different types of information. For formulating the 'forecast-objectives', we need broad generalised information, closely related to what has actually happened in the past. For formulating 'action-objectives', we need very specific, detailed information which, on the basis of a creative idea, brings to light our deficiencies and unrealised potential.

Both types of objective are needed and are therefore important. The former serves to keep things in order, while the latter serves to produce *action* which breaks through the established order. The information necessary to keep things in order can be routinised and therefore, for convenience, computerised. But the information necessary to follow up, test and implement new, creative ideas cannot be routinised and can therefore rarely be provided from computer programmes. It must be flexible and imaginative, bringing to light aspects and features previously unnoticed or overlooked, and is best produced by *ad hoc* surveys. If we think of computerising our entire information system, we are in great danger of turning management into pure administration and losing the dynamism of real entrepreneurship.

Finally, let me say a few words about information relating to the future. Naturally, every action we take today is based on some view of the future. The difficulty about the future, however, is this: that there are no facts. All facts relate to the past. Our relationship to the future can be active or passive. An active relationship to the future is called 'planning'. A passive relationship is called 'forecasting'. This is a most important distinction, albeit one that is all too often neglected. To give an example; if you ask me for future accident rates in the mining industry, I adopt a passive attitude and give you a forecast. You would think it absurd if I gave you a set of figures which I called a plan. But if you ask me to give you a programme for accident prevention, you expect me to adopt an active attitude and to give you a plan of action designed to change established patterns, and you would think it highly unsatisfactory if I merely gave you a forecast.

Planning, in other words, is the advance determination of our own actions, while forecasting is an attempt to picture what is likely to happen, without any intervention (or new intervention) on our part. As you might expect, the confusion arising from a failure to distinguish between planning and forecasting

183

is endless, like the confusion that must necessarily result when people cannot distinguish between active and passive.

If we agree, then, that planning is the advance determination of our own actions, it is quite clear that we cannot 'plan the future', because our own actions are only a very small part of the forces shaping the future. For the rest, we have to rely on forecasts relating to other people's actions. In other words, any plan we might make will depend on all sorts of forecasts about the environment into which the plan is meant to fit. While planning (in this strict sense) is relatively easy, forecasting is exceedingly difficult and often impossible.

In fact, it is far too easily assumed nowadays that the future can be predicted by means of all sorts of sophisticated techniques, data banks, and so forth. In the very short term, of course, the future is highly predictable, simply because big things take time to change. But in the longer term, only the 'broad contours' of the future landscape are predictable; the details are, as it were, still in the making, and no amount of cleverness can anticipate them. It is necessary, therefore, to be very tough and honest about forecasting. The task might be stated as follows: 'Forecast whatever, in the nature of things, is forecastable; but do not spoil your forecast by filling in details which are not, in the nature of things, forecastable.'

Having got this far, we shall want to enquire a bit more deeply into the question of 'forecastability'. Some facts or factors are stable and constant, for instance the ascertained laws of physical nature, a country's geography and climate, and certain deeply-ingrained traits of human nature. Other factors are changing, but slowly, or at ascertainable rates, such as size of population, or a country's equipment of very durable capital goods. Other factors, again, are highly indeterminate, capable of changing suddenly and unpredictably, like fashions of various sorts. Now, predictability depends on things being constant or stable or subject to change according to constant laws. If they are not constant or stable and not subject to constant laws, then, whether we like it or not, they are not predictable.

'Whether we like it or not.' Yes, and very often we do not like it. We do not like to be given a forecast which shows only the 'broad contours' of the future and misses out most of the detail which we should dearly want to know − like a map with large

184

wide areas of unexplored territory. 'Explore it!', we say. 'And if you cannot explore it, make an assumption; in fact, make as many assumptions as you need to fill in all the missing details.' And, sure enough, by slipping in the necessary number of assumptions it is always possible to produce a forecast which will captivate by its spurious verisimilitude.

The future is like a territory where we have not been before. So, a map of this territory is obviously useful. But it must be an honest and truthful map, not an imaginary one. A man who walks by an imaginary map is far more likely to go wrong than a man who has no map at all and knows that he has to rely on his wits. This is not an argument against planning, against having the best possible maps. The best possible map of the future is an honest map, and honesty demands the frank admission that there are many things – in particular, many interesting details – about the future which are not known and cannot possibly be known.

It is an unalterable law that good information is contaminated and spoiled by the admixture of spurious information.

I should beg all planners to bear this in mind all the time. It takes courage to tell superior authority, when they ask unanswerable questions: 'I do not know and I cannot see any possibility of knowing'. But this answer is often the only honest and useful answer.

It is the essence of entrepreneurial decisions that they have to be taken in the face of manifold uncertainties about the future. Indeed, if there were no such uncertainties, the decisions could be taken by clerks. Honest forecasting can often demonstrate that items which seemed highly uncertain at first sight can, in fact, be forecast with a fair degree of certainty, and it would be a dereliction of duty not to 'forecast the forecastable'. For the rest, the absence of knowledge has to be frankly admitted, and a decision taken in the full consciousness of one's lack of knowledge will certainly be a better decision than one taken on the basis of spurious forecasts derived from arbitrary assumptions.

7

From 'A question of size', *Small is Beautiful*, Blond & Briggs, 1973, pp. 58 and 59[1]

Even today, we are generally told that gigantic organisations are inescapably necessary; but when we look closely we can notice that as soon as great size has been created there is often a strenuous attempt to attain smallness within bigness. The great achievement of Mr Sloan of General Motors was to structure this gigantic firm in such a manner that it became, in fact, a federation of fairly reasonably sized firms. In the British National Coal Board, one of the biggest firms of Western Europe, something very similar was attempted under the Chairmanship of Lord Robens; strenuous efforts were made to evolve a structure which would maintain the unity of one big organisation and at the same time create the 'climate' or feeling of there being a federation of 'quasi-firms'. The monolith was transformed into a well co-ordinated assembly of lively, semi-autonomous units, each with its own drive and sense of achievement. While many theoreticians — who may not be too closely in touch with real life — are still engaging in the idolatry of large size, with practical people in the actual world there is a tremendous longing and striving to profit, if at all possible, from the convenience, humanity, and manageability of small-ness.

8

From 'The critical question of size', *Resurgence*, May–June, 1975

Statisticians tell us that the proportion of 'gainfully employed' persons in the service industries is rising while that of industrial workers is falling. This is a development with far-reaching consequences. The production of goods can be and indeed has been, handed over to machines, and this has led to the so-called growth in productivity which in turn has made possible the growth of incomes. Where do the services stand with regard to

186

the growth of productivity? Can the rendering of services be handed over to machines? The answer is an absolute No. If the human factor is taken out of the service, the service disappears and its place may or may not be taken by a labour-saving device. People's need to render service to their fellows cannot be satisfied if machines take their place. The human element disappears.

Of course, it cannot disappear altogether, and where actual people continue to render actual services — teachers, nurses, and countless others — increases in productivity cannot be generally obtained, because they mainly depend on machines, not on people. To the extent that advances in wages are made dependent on advances in productivity, the service industries tend to fall behind. But the people in the service industries, not surprisingly, insist on keeping in step with the others. As a result, the service industries' costs rise very much faster than those of other industries, and the pressure on them to 'rationalise' increases. But how can you rationalise services? Only by reducing the human element, by substituting machines, or by reducing the service. The drive for higher productivity and lower costs in the service industries therefore almost inevitably results in a further elimination of the human factor.

If my description is correct, it follows that our need to render service to our fellows is becoming more and more difficult to satisfy. The difficulty is compounded as the size of service organisations increases and as, in the pursuit of efficiency, they become centralised and more 'scientifically' organised.

The bigger an organisation, the more difficult it becomes to keep the human touch.

It strikes me as astonishing how little systematic study has been given to the all-pervading question of size. Aristotle knew about its importance, and so did Karl Marx who insisted that with changes in quantity you get at certain thresholds, changes in quality. Aristotle said: 'To the size of states there is a limit, as there is to other things, plants, animals, implements; for none of these retain their natural power when they are too large or too small, but they either wholly lose their nature or are spoiled.'

Organisations, like these 'other things', may well grow to such a size that they wholly lose their nature or are altogether spoiled. An organisation may have been set up to render various services to all sorts of helpless, needy people; it grows and grows, and suddenly you find that it does not serve the people any more but simply pushes them around. There may be complaints that the organisation has become 'too bureaucratic' and there may be denunciations of the bureaucrats. There may be demands that the 'incompetent bosses' of the organisation should be replaced by better people. But few people seem to realise that bureaucracy is a necessary and unavoidable concomitant of excessive size; that bureaucrats cannot help being bureaucrats; and that the apparent incompetence of the bosses has almost nothing to do with their personal competence.

A large organisation, to be able to function at all, requires an elaborate administrative structure. Administration is a most difficult and exacting job which can be done only by exceptionally industrious people. The administrators of a large organisation cannot deal concretely with real-life problems and situations: they have to deal with them abstractly. They cannot enjoy themselves by devising, as it were, the perfect shoe for a real foot: their task is to devise composite shoes to fit all possible feet. The variety of real life is inexhaustible, and they cannot make a special rule for every individual case. Their task is to anticipate all possible cases and to frame a minimum number of rules — a small minimum indeed! — to fit them all. We all know that life, all too often, is stranger than fiction; the dilemma of the administrators, therefore, is severe: either they make innumerable rules the enforcement of which then requires whole armies of minor officials, or they limit themselves to a few rules which then produce innumerable hard cases and absurdities calling for special treatment; every special treatment, however, constitutes a precedent which is, in effect, a new rule.

The organisation as a whole, at the same time, is faced with a further dilemma: either it draws its best brains into the administration whereupon they may be missed at operational level; or it uses its best talents at operational level, whereupon there may be serious frustration down below, owing to incompetent administration.

If there is any truth in this (very rough) analysis, the conclusion is obvious: let us organise units of such a size that their administrative requirements become minimal. In other words, let us have them on a human scale, so that the need for rules and regulations is minimised and all difficult cases can be resolved, as it were, on the spot, face to face, without creating precedents — for where there is no rule there cannot be a precedent.

The problem of administration is thus reduced to a problem of size. Small units are self-administrating in the sense that they do not require full-time administrators of exceptional ability; almost anybody can see to it that things are kept in reasonable order and everything that needs to be done is done by the right person at the right time.

I should add that, as Aristotle observed, things must be neither too big nor too small. I have no doubt that for every organisation, as for other things, there is a 'critical size' which must be attained before the organisation can have any effectiveness at all. But this is hardly a thought that needs to be specially emphasised, since everybody understands it instinctively. What does need to be emphasised is that 'critical size' is likely to be very much smaller than most people in our mass society are inclined to believe.

7
Coda

Although the force of events has compelled the adoption of ideas Fritz Schumacher started to advocate thirty years ago, he would not be easy about the energy policies now being implemented four years after his death: certainly he would not be complacent at having been proved right in his previous statements about energy.

During the twenty years we were close colleagues he often rehearsed with me material intended for his speeches and writings. If we could have such a conversation now in preparation for a talk he was to give to, say, an international conference on energy, there would be three or four main themes he would develop.

First, he would certainly show that despite current and recent political turmoil in the Middle East, OPEC's success in pushing up oil prices five-fold between January 1974 and 1977 and doubling them again in 1979, the world continues to use up those fossil fuels whose reserves are least — oil and natural gas — more than twice as fast as it is making use of coal, which is perhaps five or six times more abundant in the earth's crust.

Naturally, he would rejoice that Britain had achieved independence in terms of energy supplies; he would urge the continued expansion of the main indigenous fuel, coal, and the optimum development of renewable sources like wind power and geo-thermal energy, in an attempt to maintain self-sufficiency when North Sea oil and gas start to decline in about a decade from now.

When the price of oil began its furious gallop in 1974, the rate of growth in consumption faltered for a couple of years but then resumed its relentless progress, though at a slower rate of increase. This slowing down may be partly due to greater efficiency in its use but certainly also to the recession which itself was caused by the price of energy.

Britain, in a rough attempt to compel savings in use, has followed a high-price energy policy in a deflationary context and the recession is, perhaps partly because of this, deeper here than it is elsewhere. Industrial output is now less than at any time in the last twelve years if North Sea oil and gas are excluded. One result is that capital for energy-saving equipment and conversion to coal is difficult to raise. Therefore, Schumacher would argue for higher investment incentives so that when the economy quickens again, the effect of further oil price rises will be offset and the industrialised countries will be less dependent on the tension-ridden Middle East and less vulnerable to oil-induced inflation.[1]

He would urge the importance of a really vigorous programme for conservation with the use of combined heat and power systems, gas-fired heat pumps for space heating, fluidised bed combustion plant and integrated gasification combined cycles to improve the efficiency of power generation from coal. In the case of agriculture, where about 5 per cent of the world's annual oil supplies are being used for fuel, fertilisers and biocides, he would argue for greater use of organic farming methods.

Schumacher would disapprove the extreme financial policies of the present British Government and would criticise Ministers who, in their attacks on public expenditure, failed to make any exception for profitable investment by the nationalised industries. Public investment in energy production and research is not a luxury. Loans have to be repaid and so does the interest at a good rate of return to the taxpayer. Furthermore, since the nationalised industries undertake a big part of the country's total capital investment and buy most of the equipment and many of the services they need from the UK private sector, reductions in their programmes affect a great number of companies and their employees.

He would suggest other ways of maintaining employment. One of the economic arguments in favour of nationalisation, he would point out, is that publicly owned industries can be used to smooth out the booms and slumps in the economy. The coal industry, for example, has often put coal to stock when demand has fallen away and picked it up again two or three years later when the economy has recovered. Such a policy helps to reduce unemployment and lessen the likelihood of social strain. And since coal put to stock appreciates in value as energy prices rise, it is a valuable investment. Government grants to finance the extra stocks are sensible, therefore.

However, as a good democrat he would accept that the publicly owned energy industries have to work to the policies of the government; he would try to show how they could be managed with skill and imagination through the recession, maintaining capacity to meet demand when it starts to grow again.

He would certainly warn against becoming dependent on imports of coal which are cheap now, but which will again become expensive when world trade revives.

The most important part of his speech though, would be the promotion of an idea which could help both the oil-producing countries and the industrialised nations. He first tentatively made this proposal in a speech to the Scientific and Medical Network in May 1977 — four months before he died. The suggestion he was intending to put to the OPEC leaders, some of whom he knew, was for them to make a policy statement that they would produce each year (in other words, use up) only 2 per cent or a little more of their proved reserves. (In 1979 they supplied something like 2·75 per cent, so a threshold period of four or five years might be reasonable.)

This would avoid, for the consuming countries, a frantic scramble for supplies towards the end of the century; and the producers would spin out their resources for 100 years or so, giving them longer to develop alternative economic systems. Conservation would then be treated seriously by both groups. The process would be a good deal less violent than it would otherwise be. The co-operation between the members of OPEC would require a much greater sense of common purpose than they are presently displaying when, for example, Iran and Iraq are at war.

Finally, Schumacher would again argue the relevance of an intermediate technology for the developing countries. Costly energy has made their chances of raising their living standards infinitely more hopeless. Their need for help in turning to alternatives is greater than ever.

Biographical Note

Ernst Friedrich Schumacher was born in Bonn on August 16, 1911, the third child in the family of Hermann A. Schumacher, Professor of Economics at Berlin University. From school he went in 1929 to Bonn University, where he studied economics and political science. He was awarded the first post-war German Rhodes Scholarship and was from 1930 to 1932 at New College, Oxford. Again thanks to the Rhodes Trust he was able to go to Columbia University in New York, where he completed his course and joined the staff as a lecturer.

Schumacher took no degrees during his formal education but accepted the honorary award of Doctor of Science from the Technical University of Clausthal, West Germany, in 1963.

To gain practical experience he worked briefly between 1934 and 1936 for the Hamburg banking house of M.M. Warburg and Company, the Chase National Bank of New York, and the Reichskredit Gesellschaft A.G., Berlin. By the summer of 1934 he had decided to take a doctor's degree at a German university and, if possible, to have an academic career. He later wrote, in a curriculum vitae: 'I soon found that in view of the conditions prevailing throughout German academic life these plans had to be abandoned.'

In 1935 he joined a German organisation formed by leading industrial concerns and export houses for the purposes of helping international trade. The Nazi Government awarded Schumacher the highest German award for life saving in 1935 when he rescued a man from drowning. Although his work was congenial and his material conditions good, he decided to leave Germany and emigrate to England, which he did in January

1937, with his wife Anna Maria Petersen whom he had married the year before.

In the curriculum vitae already quoted he said '. . . I could not get used to the system which had developed in my native country. I realised, moreover, that, whatever I did, I would be contributing my part — however small — not to the wellbeing of the German people, but to the wellbeing of the Nazi war machine.'

From 1937 to 1939 he worked in investment finance in London; after a brief internment when the war began, he was employed from 1940 to 1943 as a farm labourer in Northamptonshire. In the latter year he published a paper, 'Multilateral Clearing' in *Economica,* the basic proposition of which was adapted by Professor J.M. Keynes and published as a Government White Paper, *Plan for an International Clearing Union.* This, his first public success, led to his appointment as a member of the Oxford Institute of Statistics and work with Lord Beveridge on his important and influential report, *Full Employment in a Free Society*. Thus, the former enemy alien was already contributing to social policy in the country he adopted and loved. About the same time he was writing leading articles for *The Times, The Economist* and the *Observer*. (Many years later, defining the qualities needed by an economist to make the best use of his subject, he recommended a triple foundation of: a sound theoretical knowledge of economics; a thorough experience of practical economic life; and a certain journalistic experience of how to make himself understood. His later progress abundantly justified his judgment that: 'A career built on this triple foundation could take him anywhere'.)

Schumacher became a British citizen after the war and returned to Germany, first to carry out a study for the Americans of the effects of strategic bombing and then in 1946, when he was appointed Economic Adviser to the Allied Control Commission for Germany, working in Berlin and Frankfurt: one of the main tasks of this organisation was the restoration of the German coal industry.

In 1950 Schumacher began his twenty-year service as Economic Adviser to the National Coal Board, with the added responsibility of Director of Statistics from 1963, and of Head of Planning four years later. Twice during his time at the NCB he was seconded to advise overseas governments — that of Burma in 1955 and of India in 1962. These experiences saw the beginning of his concern for the people of the developing countries and led him to found in 1965 the Intermediate Technology Development Group, through which he worked to provide aid and advice of a practical kind suited to the needs and resources of each of the many countries with which he was involved until his death in September 1977.

Small is Beautiful (Blond & Briggs) was published in 1973 and *A Guide for the Perplexed* (Jonathan Cape) a few weeks after he died. *Good Work* (Jonathan Cape), based on lectures he gave during his last tour of North America, appeared in 1979, and *Small is Possible* by George McRobie (Jonathan Cape) came out in 1981.

Schumacher was a director and trustee of a common ownership firm, Scott Bader and Company Limited, and from 1970, President of the Soil Association of Great Britain. In 1974 he was awarded the CBE by the British Government.

His first wife died in 1960 leaving two sons and two daughters; he and his second wife, the Swiss-born Verena Rosenberger, also had two sons and two daughters.

Notes

The numbered notes have been compiled by the Editor. Those indicated by asterisks, etc., are Schumacher's own.

Editor's Introduction

1 Jonathan Cape, 1977
2 Blond & Briggs, 1973
3 Cassell, 1972

Chapter 1 *The Principles of Energy Conservation*

EXTRACT 1

1 *Statistical note:* There are three different units of weight in this volume: the British ton; the metric tonne (about 1·6 per cent smaller); and the US short ton (10 per cent smaller still). The metric tonne was adopted in the UK in 1978.

EXTRACT 3

1 There is more material from this paper on pp. 61–3 and 79.
2 In *The Coal Question,* 1865.
3 Oxford Professor of Thermodynamics and, at the time of his death in 1956, Head of the Clarendon Laboratory in Oxford.

EXTRACT 4

* What to take now and what to leave for later is, in fact, a question of ethics, not of economics.

1   Sir James Bowman who was also formerly Vice-President of the National Union of Mineworkers.

EXTRACT 5
1   More material from this paper is on pp. 144–6.

EXTRACT 7
1   This is from the English version written by Schumacher.

EXTRACT 8
1   This extract is from the earlier part of the lecture. Schumacher used the latter part as the basis for the chapter 'Nuclear Energy – Salvation or Damnation' in *Small is Beautiful*.
*   Cf. *The Financial and Economic Obligations of the Nationalised Industries,* Cmnd 1337, HMSO, London, 1961.
†   *Time,* the weekly news magazine, August 26, 1966, p. 11.
‡   John Kenneth Galbraith, *The Affluent Society,* Penguin Books, 1962, p. 211.
§   US Public Health Service, *Health Statistics from the US National Health Survey, series C, No. 5,* Washington, DC, 1961, p. 1.
** Ralph and Mildred Buchsbaum, *Basic Ecology,* Pittsburgh, 1957, p. 20.

EXTRACT 9
1   From which Extract no. 8 was taken. The passages which caused most official anger were those relating to the health hazards of nuclear power and were used by Schumacher in *Small is Beautiful*.
2   Central Electricity Generating Board.

EXTRACT 10
1   District heating is a method of supplying heating and hot water from a central boiler-house to a number of adjacent users – for example, dwellings, offices, factories, hotels, schools and shops. The film was an attempt to promote its use.

EXTRACT 11
1 Another extract from this article is on pp. 48–53.

EXTRACT 12
* *Surveys on Energy and Lifestyles*, 1972–3.

Chapter 2 *The World's Energy Needs and Resources*

INTRODUCTION
1 The Reagan administration, coming into office in 1981, set out to reduce the size of this ambitious United States' programme.

EXTRACT 1
1 Another short extract from this address appears on p. 58.
2 World population in fact grew much faster than expected and in 1978 was 4,258m.
3 Or 10 million therms.
4 In 1979 only 2·5 per cent of world energy was from nuclear power.
5 *Coal — Bridge to the Future, The World Coal Study (WOCOL)*, Ballinger Publishing Company, Cambridge, Massachusetts, May 1980, suggests that the world trade in steam coal will have to grow ten to fifteen times in the next twenty years.

EXTRACT 2
1 A longer extract from this paper was reproduced in *Small is Beautiful*, Blond & Briggs, 1973, pp. 116–20.
2 Oil consumption in 1979 (latest available figures) in fact was 3,120m. tonnes, which was indeed its highest ever level, having increased by about 2·5 per cent a year since 1975. Reserves in the same year were nearly 90,000m. tonnes — again the highest recorded level, but representing only about 28 years' life at the current annual consumption.
3 Ibid.

EXTRACT 3

1 In the event, coal production in the United States in 1975 was 655m. short tons.
2 Total world consumption of oil in 1962 was 1,216m. tons. By 1979, it had risen to 3,222m. tonnes and was increasing by 3 to 4 per cent a year, so the twenty-year growth is likely to be three-fold, not four. The recession that followed the Arab-Israeli War in 1973 severely affected oil use. (Figures from *BP Statistical Review of the World Oil Industry.*)

EXTRACT 4

1 See also pp. 91–6 for other material from this paper.
2 By 1979 output had reached 747m. tonnes.
3 It was already 10,440 million tons coal equivalent by 1979.
4 In late 1979 the life of published proved reserves was down to 28 years.

EXTRACT 5

1 For another extract from this speech see pp. 96–8.

EXTRACT 6

1 There is more material from this speech on pp. 98–101.
2 It turned out to be 17 per cent and in 1979 reached 20 per cent.

EXTRACT 7

1 More material from the article appears on pp. 23–5.
* *CBS Notes* (Center for the Biology of Natural Systems, Washington University, St Louis, Missouri), Volume 6, 1, 5 (1973).
† B. Commoner, *The Closing Circle,* pp. 149–50, Alfred A. Knopf, New York, 1971.
‡ Ibid.
§ *Pollution: Nuisance or Nemesis?* p. 36, HMSO, London, 1972.
** *Sinews for Survival,* pp. 37–8, HMSO, London, 1972.
†† Ibid., p. 36.
‡‡ Ibid.

EXTRACT 8
1   See also pp. 74–5 and 106–7 for further extracts from this address.

Chapter 3 *Exploratory Calculations into Energy Consumption*

EXTRACT 1
1   Two short extracts from this review were reproduced in *Small is Beautiful*, Blond & Briggs, 1973, pp. 114–15.

EXTRACT 2
1   See pp. 31–8 for more material from this address.

EXTRACT 3
*   H. Brown, J. Bonner and J. Weir, *The Next Hundred Years*, Weidenfeld & Nicolson, 1957.

EXTRACT 4
1   Other brief extracts from this paper are on pp. 5–6 and 79.
2   Jevons's own italics.

EXTRACT 5
1   With more than twenty years to go (in 1979), world total energy consumption had already reached 10,440 million tons of coal equivalent.
2   Schumacher reproduced here the table which appears on p. 3 of the present volume.

EXTRACT 6
1   Another extract from this article is on pp. 89–91.

EXTRACT 7
1   Pages 101–3 also contain material from this address.
*   *Report of the Committee on National Policy for the use of Fuel and Power Resources*, HMSO, Cmnd 8647, September 1952.
2   *U.S. Energy Policies: an agenda for research*, distributed by Johns Hopkins Press, Baltimore, 1968.
3   The theme of the rest of this extract was later developed in *Small is Beautiful*, Blond & Briggs, 1973, pp. 21–4.

EXTRACT 8
1   See also pp. 53 and 106–7 for other extracts from this
    Paper.
2   HMSO, 1967, Cmnd 3438.

Chapter 4 *The Causes of Crisis — Shutting Down the Mines*

INTRODUCTION
1   *Meeting Europe's Energy Requirements,* 1963; *An Energy
    Policy for Europe,* 1966; and *Energy Resources for Western
    Europe,* 1970.
2   HMSO, Cmnd 3438, 1967.
3   See Lord Robens, *Ten Year Stint,* Cassell, 1972, chapter 9,
    for the Coal Board's attitude to nuclear energy during
    Robens's Chairmanship.
4   *Nuclear Power and the Energy Crisis,* The Macmillan Press,
    for the Trade Policy Research Centre, 1978.

EXTRACT 1
1   There are longer passages from this paper on pp. 5–6 and
    61–3.

EXTRACT 2
1   January 1, 1947 — when the industry passed into public
    ownership.
2   In 1981 the crises in the West European coal and steel
    industries have a different cause — over-production in a
    time of severe recession.
3   A phrase used by a writer in the *Guardian*.

EXTRACT 4
1   *Towards a new energy pattern in Europe: Report of the
    Energy Advisory Commission of the Organisation for Euro-
    pean Economic Co-operation,* Paris, January 1960.

EXTRACT 6
1   *Meeting Europe's Energy Requirements*. Published by the
    National Coal Board and the West European Coal Produc-
    ers, February 1963.

EXTRACT 8
1 Other material from this article is on pp. 67–8.

EXTRACT 9
1 See also pp. 43–6 for other material from this paper.
2 In 1979 (latest available figures) the total, excluding the
   Eastern bloc, was 110,000 Mw, equivalent to 262m. tonnes
   of coal.
3 *Civilian Nuclear Power,* a report to the President, US
   Atomic Energy Commission, Washington, November
   1962.
4 In 1979 it was about 2·5 per cent.
5 At the end of 1979 the Middle East share of world published
   proved reserves was 56 per cent.
6 Between 1965 and 1978 the increase was 58 per cent.

EXTRACT 10
1 There is another extract from this speech on p. 47.
2 For the actual size and importance of North Sea reserves,
   see p. 30.
3 The increase turned out to be much smaller — from 303
   million tons coal equivalent to 325m.

EXTRACT 11
1 See p. 48 for another extract from this article.
2 This expectation proved extremely over-optimistic: by
   1955 it was only 11 million tons coal equivalent and even in
   1980, nuclear power's contribution was still only about
   14 m.t.c.e.

EXTRACT 12
1 See also pp. 68–73 for more extracts from this address.
2 *Fuel Policy,* HMSO, Cmnd 3438, November 1967.
3 The estimate for nuclear power and hydro electricity
   indeed proved to be over-optimistic: their combined contribution in 1970 was only 11·6m.t.c.e. Natural gas did
   slightly better than expected, supplying 17·6m.t.c.e. from
   all sources.
4 In the British coal industry.

EXTRACT 13
1 Two more quotations from *Small is Beautiful* will be found on pp. 148–51 and 186.
2 Of the kind quoted many times in the present volume.
3 HMSO, Cmnd 3438
4 *Pollution: Nuisance or Nemesis? Working Party Report on The Control of Pollution,* HMSO, 1972.

EXTRACT 14
1 See also pp. 53 and 74–5 for other extracts from this paper.

Chapter 5 *The Principles of Public Ownership*

INTRODUCTION
1 Vandenhoeck and Ruprecht, Göttingen, 1962.

EXTRACT 1
1 See Introduction to Chapter 5, note 1.
* William A. Robson, *Nationalised Industry and Public Ownership,* Allen & Unwin, 1960, p. 77.
2 The Lord Chief Justice, Lord Goddard.
3 The Rt Hon. Herbert Morrison, MP, later Lord Morrison, who held many Ministerial posts in the war-time Coalition Cabinet and the Labour Government of 1945–51 when he was Deputy Prime Minister, Lord President of the Council and Leader of the House of Commons.

EXTRACT 6
1 Souter is the Scots word for a shoemaker.

EXTRACT 8
1 There is another extract from this paper on pp. 10–13.

EXTRACT 9
1 National Economic Development Organisation.

EXTRACT 10
1 For further extracts see pp. 104–6 and 186.

2   R.H. Tawney, *The Acquisitive Society,* Bell, 1952.
3   Ibid.

Chapter 6 *Making the Figures Sing — Business Management and the Use of Statistics*

EXTRACT 1
1   Headquarters of the National Coal Board.

EXTRACT 3
1   In decimal currency, a little over 3.5p out of £2.80.

EXTRACT 7
1   Two other extracts from *Small is Beautiful* appear on pp. 104–6 and 148–51

Chapter 7 *Coda*

1   The British Government's decision in March 1981 to provide grants up to a total of £50m. in two years for conversion of oil- to coal-burning equipment is a good beginning but is hardly on a sufficiently big scale.

Index

208

209

Reichskredit Gesellschaft AG,
Berlin, 194
Resources: renewable, 38, 63;
non-renewable, 6, 11, 12, 63,
108, 144
Resources for the Future, Inc., 71,
73
Rhodes Scholarship, 194
Ridley Committee, 68
Robens of Woldingham, Lord:
attitude to NCB officials, xiii:
comments on nuclear power,
78n, Government attitude to
coal, xvi, xvii; leadership, 87;
opposition to Governments, 77;
reorganisation of NCB, 153,
186; sales objective, 88; *Ten
Year Stint,* xvi
Robinson Report, *see* Organisation
for European Economic
Co-operation
Robson, Professor William A.,
*Nationalised Industry and Public
Ownership,* 116
Roman Catholic faith, xi
Rosenberger, Verena, 196

Saharan oilfield, xiv, 14, 39, 104
Samson Stripper, 174
Saudi Arabia, 102, 109
Schumacher, E.F.: biography,
194–6; skill as communicator,
xiii–xiv; use of English, xiii
Schumacher, Professor Hermann
A., 194
Scott Bader and Company
Limited, 196
Selby coalfield, 54
Shale oil, 38
Simon, Sir Francis, 5n
Smith, Adam, 18
Socialism, British aims and
history, 114–15
Socrates, 174
Soil Association of Great Britain,
196
Solar energy, 38, 50–1, 79

Statistics: coalfaces, 152, 157–9;
colliery profile, 156;
management needs, 176–9;
measuring efficiency, 154–5;
nature of, 152, 154; Parkinson's
Law, 177–8
Steel industry: crisis, 81;
nationalisation, 112; world
expansion, 43
Strikes, 54, 78, 108, 131
Synthetic fuels: future needs, 35;
research, 31; *see also*
Liquefaction of coal, *and*
Gasification of coal

Tar sands, 38, 109, 110
Tawney, Professor R.H., 114, 149
Thermal efficiency, *see* Fuel
efficiency
Third World: appropriate
technology, 24, 29, 192–3, 196;
energy costs, 27–8, 192–3;
migration to cities, 24
Tidal and wave power, 38, 52
Trades Union Congress, 112

Unemployment, 191
University mining departments,
88
Uranium, 94, 95
United States: Arab investment,
109–10; Atomic Energy
Commission, 65; Bureau of
Mines, 28, forecast of 1975 coal
output, 42n, study of *fuel cost* of
fuel, 28; coal, exports to Europe,
55, output forecasts, 42, 56,
pithead price, 56; Health Service
report (1961), 20; Nuclear
Energy Authority, 101

Vegetable sources of power, 38

Warburg, M.M. and Company,
194
Washington Center for
Metropolitan Studies, 25

211

SMALL IS
POSSIBLE
George McRobie

The sequel to E. F. Schumacher's
SMALL IS BEAUTIFUL
and
A GUIDE FOR THE PERPLEXED.

The ballast of the wealth of nations has shifted since the late E. F. Schumacher first claimed that small is beautiful, stressing the worldwide human and financial need for 'economics as if people mattered'. The oil crisis predicted by Schumacher has sliced into Western industrial investment. At home and abroad business corporations whose gigantic bulk was mistaken for muscle have staggered to the wall, dinosaur victims of economic evolution. Now more than ever we need to realise the potential of alternative technology and invest in a different tomorrow. In SMALL IS POSSIBLE Schumacher's friend and collaborator George McRobie gives an optimistic progress report from the front line and proves that there *is* an alternative future for all of us.

ECONOMICS   0 349 12307 1   £3.25

THE
SCHUMACHER LECTURES

Edited with an introduction by Satish Kumar

With contributions from:

LEOPOLD KOHR	HAZEL HENDERSON	R. D. LAING
EDWARD de BONO	AMORY LOVINS	IVAN ILLICH
JOHN MICHELL	FRITJOF CAPRA	

The Schumacher Lectures embraces economics, psychology, physics, linguistics, history and philosophy and brings together eight remarkable and original thinkers with an astonishing range of ideas, which stand as a testimony to the immense influence of E. F. Schumacher's work. The contributors share a common conviction that our society can no longer accept the demands of technology and 'progress' – we must actively determine our destiny and our own real needs.

Satish Kumar is the Chairman of the Schumacher Society and Editor of *Resurgence*, with which Dr Schumacher was closely associated. After Dr Schumacher's death in 1977, *Resurgence* launched the Schumacher Society, which now holds an annual series of lectures in honour of the bestselling author of *Small is Beautiful*.

'A thought provoking and sometimes inspiring taste of the current ideas of some influential thinkers, it is to be recommended.' *Time Out*

PHILOSOPHY   0 349 12118 4   £2.50

GOOD WORK

BY E. F. SCHUMACHER

WHAT IS THE PURPOSE OF OUR WORK?

When Dr Schumacher took a look at the plight of businesses today and said 'small is beautiful' he spoke to millions and coined a phrase. In GOOD WORK he addresses a question which is central to most of us and one which is all too often ignored by the economic structure of the Western world. Dr Schumacher maintains that the purpose of man's work is threefold: to produce necessary, useful goods and services; to enable us to use and perfect our gifts and skills and, finally, to serve and collaborate with other people in order to liberate ourselves from inbuilt egocentricity. A job in which one finds no personal satisfaction destroys the soul. With sanity and sensitivity the late E. F. Schumacher offers important and thought-provoking alternatives which point the way to mankind's physical and mental liberation.

'Compulsive reading'
TIME OUT

'The message of this set of essays is beguiling; technology is not the inevitable ruler and real alternatives seem possible'
NEW SCIENTIST

ECONOMICS 0 349 13133 3 £1.95

and don't miss
SMALL IS BEAUTIFUL
A GUIDE FOR THE PERPLEXED
also by E. F. Schumacher in Abacus

THE
LEAN YEARS

BY RICHARD BARNET

THE LEAN YEARS is the first book to take an overall look at the world's resources – how much we have got, where they are, and who controls them. Richard Barnet, the distinguished economic and political analyst, unravels the scarcity puzzle, exploring the politics of the 'petroleum economy', revealing the role of the giant oil companies and evaluating the energy alternatives to OPEC oil. He examines the changing face of global power and the struggle for food, water and mineral resources which we can expect in the coming lean years, and reports on the role of the multinational corporations in the world employment crisis. In addition, he discusses the politics of a world in transition, the consequences of the growing industrialisation of labour worldwide, and the absurdities of scarcity in a world of plenty. And, finally, he offers positive proposals for reorganising global resource systems to meet the needs of Earth's burgeoning population.

CONSERVATION/ECONOMICS   0 349 10238 4   £2.50

Also available from ABACUS

FICTION

SILVER'S CITY	Maurice Leitch	£2.50 ☐
SUNDAY BEST	Bernice Rubens	£2.50 ☐
THE TRIAL OF FATHER DILLINGHAM	John Broderick	£2.50 ☐
SOUR SWEET	Timothy Mo	£2.95 ☐
THE SIDMOUTH LETTERS	Jane Gardam	£1.75 ☐
GOOD BEHAVIOUR	Molly Keane	£2.95 ☐
CUSTOMS	Lisa Zeidner	£2.95 ☐
WHEN THE EMPEROR DIES	Mason McCann Smith	£3.95 ☐

NON-FICTION

PHYSICS AS METAPHOR	Dr. Roger Jones	£3.50 ☐
THE DRAGON AND THE BEAR	Philip Short	£4.95 ☐
THE SECOND STAGE	Betty Friedan	£2.95 ☐
MEDIATIONS	Martin Esslin	£2.95 ☐
THE BAD BOHEMIAN	Sir Cecil Parrott	£2.95 ☐
SMALL IS BEAUTIFUL	E. F. Schumacher	£2.50 ☐
GANDHI – A MEMOIR	William L. Shirer	£1.75 ☐
FROM BAUHAUS TO OUR HOUSE	Tom Wolfe	£2.95 ☐

All Abacus books are available at your local bookshop or newsagent, or can be ordered direct from the publisher. Just tick the titles you want and fill in the form below.

Name_____

Address _____

Write to Abacus Books, Cash Sales Department, P.O. Box 11, Falmouth, Cornwall TR10 9EN

Please enclose cheque or postal order to the value of the cover price plus:

UK: 45p for the first book plus 20p for the second book and 14p for each additional book ordered to a maximum charge of £1.63.

OVERSEAS: 75p for the first book plus 21p per copy for each additional book.

BFPO & EIRE: 45p for the first book, 20p for the second book plus 14p per copy for the next 7 books, thereafter 8p per book.

Abacus Books reserve the right to show new retail prices on covers which may differ from those previously advertised in the text or elsewhere, and to increase postal rates in accordance with the PO.